INTRODUCTION TO HYPOPLASTICITY

ADVANCES IN GEOTECHNICAL ENGINEERING AND TUNNELLING

1

Introduction to Hypoplasticity

D. KOLYMBAS

University of Innsbruck, Institute of Geotechnics and Tunnelling

CRC Press
Taylor & Francis Group
Boca Raton London New York

CRC Press is an imprint of the
Taylor & Francis Group, an **informa** business

A BALKEMA BOOK

Published by:
CRC Press/Balkema
Schipholweg 107C, 2316 XC Leiden, The Netherlands

© 2000 by Taylor & Francis Group, LLC
CRC Press/Balkema is an imprint of Taylor & Francis Group, an informa business

No claim to original U.S. Government works

ISBN-13: 978-90-5809-305-9 (hbk)
ISBN-13: 978-90-5809-306-6 (pbk)

Visit the Taylor & Francis Web site at
http://www.taylorandfrancis.com

and the CRC Press Web site at
http://www.crcpress.com

ISSN 1566-6182

Contents

Foreword of the Editor

This booklet inaugurates the series *Advances in Geotechnical Engineering and Tunnelling*. Subsequent issues will follow in irregular terms, either in German or in English language.

The present publication aims to give the reader a short, tractable and as far as possible complete introduction to the young theory of hypoplasticity, which is a new approach to constitutive modelling of granular media in terms of rational continuum mechanics.

I wish to thank Dr. I. Herle and Dr. W. Fellin for many valuable suggestions, Josef Wopfner and Christoph Bliem for working out the exercises and also Marlies Span for the thorough typing.

D. Kolymbas
Institute of Geotechnics and Tunnelling
University of Innsbruck
August, 1999

Chapter 1

Simple questions and answers

1.1 What is a constitutive equation?

A constitutive equation is a mathematical relation connecting stress and strain for a particular material. Of course, stress and strain are tensorial quantities. Besides stress and strain, some additional quantities, the material constants (e.g. YOUNG's modulus), appear within a constitutive equation. The values of the material constants adjust the constitutive equation to a particular material, i.e. they make possible to distinguish e.g. between an elastic rubber and an elastic steel.

1.2 What for is a constitutive equation useful?

In order to predict the deformations and/or the stability of a loaded body you need to know its constitutive equation. E.g., the constitutive equation of soil is needed to predict the stability of a slope or a cut, or to predict the loads exerted to the lining of a tunnel or to a basement, and also to predict the deformations around an excavation or the settlement due to tunnelling or due to extraction of oil from the underground. To answer these questions we use the balance laws of mechanics (balance of mass and momentum). Since these equations are in most cases insufficient to solve the problem, we need some additional information, which is provided by the constitutive equation.

The knowledge of the constitutive equation is necessary but not sufficient to answer the above stated questions. We also need to carry out complicated numerical calculations, mainly following the method of finite elements.

A sound constitutive equation also helps to understand the behaviour of a material. Virtually, this understanding is only possible within the framework of a constitutive equation.

1.3   What is elasticity?

The property of elasticity is given if the stress (or the strain) depends uniquely on the strain (or the stress). This means that the strain (or stress) *history* is immaterial and only the *actual* value of the strain (or stress) is needed to determine the actual value of stress (or strain). This property is also called path-independence, as the previous history can be conceived as a strain- (or stress-) path. In mathematical terms, elasticity means that the stress is a *function* of strain, or — vice versa — the strain is a function of stress. Elastic materials do not exhibit irreversible deformations, i.e. if we remove the load, the deformation (connected with this load) completely rebounds. The particular case of *isotropic* and *linear* elasticity is mathematically described by the constitutive equation of HOOKE.

1.4   Why is the theory of elasticity inappropriate to describe the behaviour of soil?

The ability to undergo irreversible deformations means the ability to memorize previous loading. For soil this memory is evident: If we walk on a sandy beach we leave traces behind. The sand is compressed by our own weight and this compression does not rebound when we unload it. Besides this very basic phenomenon there are also other important effects which cannot be described in the realm of elasticity: (i) plastic yield, i.e. the unlimited growth of deformation under constant stress, (ii) dilatancy-contractancy, which can be described as a tendency of a material to change its volume under shear deformation, (iii) stress dependent stiffness.

1.5   How can we describe anelastic (i.e. irreversible) deformations?

A constitutive equation capable of describing anelastic behaviour should manage, in some way, to provide different stiffnesses for loading and unloading. Of course, this should be accompanied by a criterion of what is loading and what is unloading. The most widespread mathematical framework for irreversible deformations is the so-called elastoplasticity. From the basic concept of elastoplasticity emanate many different models, the multiplicity of which can hardly be over-viewed. Many scientists believe that elastoplasticity is the only framework to describe anelastic materials. They ignore that there is a alternative to elastoplasticity given by the young branch of hypoplasticity.

1.6 What is elastoplasticity?

According to elastoplasticity, a material behaves in the initial stage of deformation elastically, whereas plastic deformations set on later in the course of a continued loading. The onset of plastic deformations is determined by a surface in stress space, which is called the yield surface. The direction of plastic deformations is determined by another surface, the so-called plastic potential, whereas its magnitude can be determined from the so-called consistency condition, which requires that a stress point carries behind the yield surface, when the material is loaded. Thus, elastoplasticity is characterized by a series of additional notions (mainly of geometric nature) which hide the mathematical structure of the constitutive equation. The various elastoplastic constitutive equations are altogether hardly tractable, difficult to be implemented in FEM-codes and extremely sensitive to parameters controlling the various involved numerical algorithms. Usually, such disadvantages are neither remarked nor confessed.

1.7 What is hypoplasticity?

Hypoplasticity aims to describe the aforementioned anelastic phenomena without using the additional notions introduced by elastoplasticity (such as yield surface, plastic potential etc.). Hypoplasticity recognizes that anelastic deformations may set on from the very beginning of the loading process. It does not a priori distinguish between elastic and plastic deformations. The outstanding feature of hypoplasticity is its simplicity: Not only it avoids the aforementioned additional notions but it also uses a unique equation (contrary to elastoplasticity) which holds equally for loading *and* unloading. The distinction between loading and unloading is automatically accomplished by the equation itself. Besides the indispensable quantities "stress" and "strain" (and their time rates) only some material constants appear in the hypoplastic equation.

1.8 What are the advantages of hypoplasticity?

There is no method to measure the success or the utility of a constitutive equation. Compared with persons, there is no way to say that person A is better than person B, even if A is an Olympia winner. However, researchers familiar with hypoplasticity find that it is easier to be implemented into numerical algorithms and is also easier to be grasped.

1.9   What does the hypoplastic constitutive equation?

The hypoplastic constitutive equation expresses the *stress increment* as a function of a given *strain increment* and of the actual *stress* and *void ratio*. Instead of stress and strain increments we can speak of stress and strain *rates*. We can conceive e.g. the stress rate as a stress increment obtained within a time unit. As stress and strain are tensorial quantities, the hypoplastic equation is a tensorial equation. Herein, the stress is denoted symbolically by $\mathbf{T}$ and the rate of strain is denoted by $\mathbf{D}$. Alternatively, the complete stress tensor could be written, i.e.

$$\begin{pmatrix} \sigma_{11} & \sigma_{12} & \sigma_{13} \\ \sigma_{21} & \sigma_{22} & \sigma_{23} \\ \sigma_{31} & \sigma_{32} & \sigma_{33} \end{pmatrix} \quad ,$$

or denoted by a representative element of this matrix, i.e. σ_{ij} (so-called index notation). However, the symbolic notation is simpler.

1.10   Why are there several versions of hypoplastic equations?

As with every constitutive equation, there are several versions of hypoplastic equations, early ones and more advanced ones. The original hypoplastic equation published by the author in 1977 (at those days the name 'hypoplasticity' was not yet launched) proved to have several shortcomings. At that, it was too complex. Later on improved versions have been introduced by several authors. Thus, hypoplasticity should rather be conceived as a *frame* of constitutive equations than a particular one. After all, a constitutive equation is not a theorem with absolute validity like e.g. the mass balance equation. A constitutive equation is a behavioural equation, as it describes the mechanical behaviour of a particular class of materials. This description is approximate and, thus, every constitutive equation can (at least in principle) be improved.

It is natural that each author who introduces a new version of a hypoplastic equation brings in his personal point of view. In such situations one could be tempted to ask: Is the version x still a hypoplastic equation or not? Or: What are the real roots of hypoplasticity? The author believes, however, that there is no use in any dogmatism.

1.11   What is the range of validity of hypoplasticity?

The present hypoplastic versions can be recommended for granular materials consisting of not too soft grains. The loading processes can comprise loading and unloading but not cyclic loading. A small cohesion can be comprised, but heavily overconsolidated soils are still out of the scope. The viscosity of the granular skeleton is not taken into account by the current versions of hypoplasticity[1]. Of course, it is hoped that improved versions in the future will cover also the above stated limitations.

---

[1]There have been several attempts to comprise rate dependence in hypoplasticity using a constitutive relation $\mathbf{h}(\mathbf{T}, \mathbf{D})$ which is not homogeneous of the first degree with respect to $\mathbf{D}$ [29, 85, 16]. A review of these relations together with a promising proposal is given by NIEMUNIS [50]

Chapter 2

Experimental results on soil behaviour

In this chapter are summarized the main experimental results that characterize the behaviour of granular materials such as sand.

2.1 Triaxial test

The triaxial test has been introduced into rock mechanics in 1911 by VON KÁRMÁN and into soil mechanics in 1928 by EHRENBERG. A cylindrical soil specimen is loaded by the stress component σ_1 in axial direction and by $\sigma_2 = \sigma_3$ in lateral directions. The corresponding strain components are ε_1 and $\varepsilon_2 = \varepsilon_3$ in axial and radial directions, respectively. We have thus axisymmetric conditions. The lateral stress is exerted by fluid (water or air) pressure acting upon the specimen. A typical load program is to increase σ_1 while keeping $\sigma_2 = \sigma_3$ constant. Another possibility is to increase σ_1 keeping the sum $\sigma_1 + \sigma_2 + \sigma_3$ constant. In principle, stress paths of arbitrary directions in a σ_1-σ_2-space can be accomplished, as long as they remain within the so-called limit surface (see Fig. 2.1). The aforementioned loading processes usually start from hydrostatic stress states. The radial strain $\varepsilon_2(= \varepsilon_3)$, or equivalently the volumetric strain $\varepsilon_v = \varepsilon_1 + \varepsilon_2 + \varepsilon_3$ is registered by measuring the volume of porewater that is expelled from a water-saturated sample during the deformation.

The measured stress is usually plotted over the axial strain $\varepsilon_1 = \Delta h/h_0$, with h_0 being the initial height of the sample. An alternative strain measure is the logarithmic strain ϵ_1 defined as $\log(h/h_0)$. For small strains the difference between ε_1 and ϵ_1 is negligible. Compressional strains are considered as negative quantities but, for traditional reasons, ε_1 is represented positive in plots. Typical stress measures are usually the deviatoric stress $\sigma_1 - \sigma_2$, the stress ratio σ_1/σ_2 and the ratio $(\sigma_1 - \sigma_2)/(\sigma_1 + \sigma_2)$. The shape of the stress-strain curves depends on the stress measure depicted in the y-axis (see Fig. 2.2). The maximum σ_1 value is denoted as peak and serves to determine the friction angle of a cohesionless material (e.g. sand):

$$\varphi := \arcsin\left(\frac{\sigma_1 - \sigma_2}{\sigma_1 + \sigma_2}\right)_{\max}$$

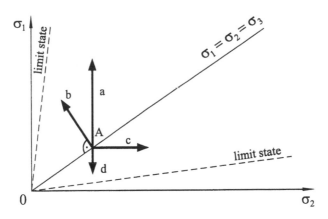

Figure 2.1: Stress paths realized by the triaxial test

Using loose samples a peak is not obtained, the stress-strain curve is monotonically increasing. In such cases the friction angle is obtained by convention from the stress at, say, $\varepsilon_1 = 10\%$. An analytical expression to represent the measured stress-strain-curves can hardly be found. The hyperbolic equation of COX (often attributed to KONDNER)

$$\sigma_1 - \sigma_2 = \frac{\varepsilon_1}{a + b\,\varepsilon_1}$$

has the disadvantage that it does not exhibit a peak.

2.1.1 Barotropy and pyknotropy

If we compare two triaxial tests carried out at different lateral stresses, we observe that the higher lateral stress is connected with a higher σ_{1max}-value. If we compare the *normalized* stress-strain curves, i.e. plots of σ_1/σ_2 or $(\sigma_1 - \sigma_2)/(\sigma_1 + \sigma_2)$ over ε_1, we observe that both curves (i.e. the one obtained at low cell pressure σ_2 and the one obtained at high σ_2) more or less coincide. This implies that the friction angle φ is independent of stress level and the tangential stiffness $d\sigma_1/d\varepsilon_1$ (taken for a fixed ε_1-value) is proportional to the stress level (i.e. to σ_2). A more thorough look reveals, however, that φ decreases with increasing stress level and the stiffness $d\sigma_1/d\varepsilon_1$ is proportional to σ_2^n with $0 < n < 1$. Also the volumetric behaviour, i.e. the ε_v versus ε_1 curve is affected by the stress level. This influence of stress level (see Fig. 2.3) is called *barotropy*.

One of the most interesting properties of granular media is that they can be encountered at various densities (i.e. various grain arrangements). Dense granulates have a higher density and a higher friction angle than loose ones and they have the tendency

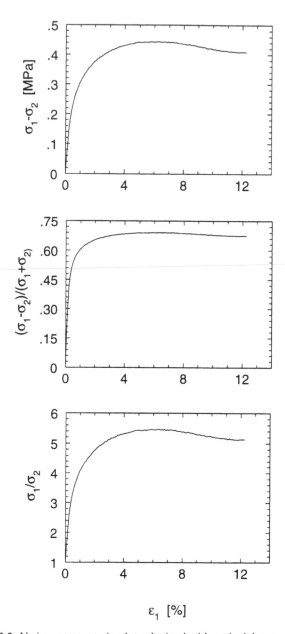

Figure 2.2: Various stress-strain plots obtained with a triaxial test on sand

to increase their volume (i.e. to decrease their density) under shear (see Fig. 2.4). This effect is called dilatancy. Loose granulates have the tendency to decrease their volume (i.e. to increase their density) under shear, which is called contractancy. The term *pyknotropy* denotes this influence of density.

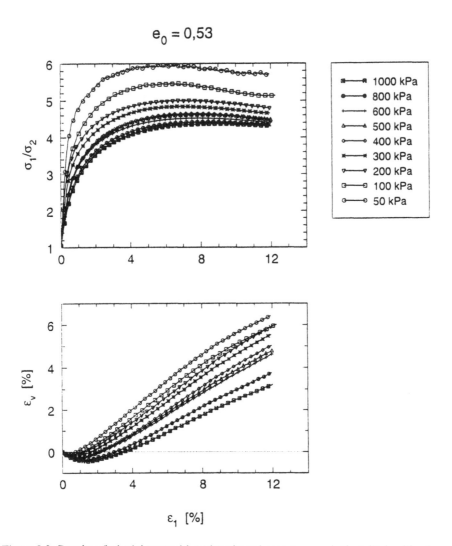

Figure 2.3: Results of triaxial tests with various lateral pressures. e_0 is the initial void ratio.

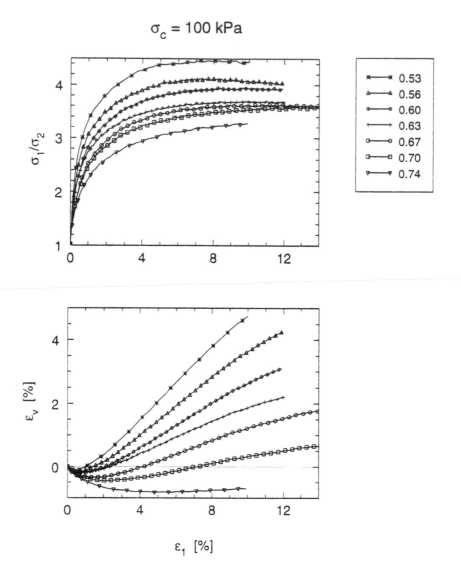

Figure 2.4: Results of triaxial tests with various initial void ratios. $\sigma_c = \sigma_2$ is the lateral stress.

2.1.2 Critical state

The aforementioned volume (or density) changes under shear (dilatancy and contractancy) are not limitless. They are bounded by the so-called critical density which
is asymptotically obtained upon continuation of shear. The critical density increases
with increasing stress level. The decrease of density is connected with a decrease of
the friction angle from peak to the so-called residual value.

2.1.3 Unloading

A cycle of loading and unloading leaves always an irreversible or plastic deformation behind (see Fig.2.5). In some materials (e.g. steel) plastic deformations are only observed if the stress exceeds a particular limit. This is, however, not the case with granular media, where plastic deformations are always obtained, no matter how small the applied stress is. Plastic deformation is the main feature to distinguish between elastic and plastic (or anelastic) materials. It is inherently connected to the fact that the incremental (or tangential) stiffness at unloading is always larger than the one at loading. This fact constitutes the so-called incremental non-linearity.

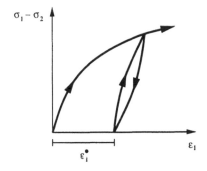

Figure 2.5: Loading, unloading and reloading

Every additional loading-unloading cycle produces an additional plastic deformation and the question is whether the sum of incremental plastic deformations is bounded (so called shake-down) or limitless (so called incremental collapse). The answer depends on many parameters and is not clear yet.

2.1.4 Homogeneity of deformation

The measured quantities in the triaxial test are forces and displacements. To evaluate the test in terms of stresses and strains we have to assume that the deformation is homogeneous, i.e. every material point of the sample undergoes the same deformation. In reality the deformation becomes increasingly inhomogeneous in the course of a triaxial test (see Fig. 2.6). Possible reasons are the self weight of the sample, which induces an inhomogeneous stress field, and the friction at the end plates. Counter measures[1], as the lubrication of end plates (to eliminate the friction), do not help completely. At that, lubrication induces an error in the measurement of axial strain, the so-called bedding error. The present state of knowledge is that inhomogeneities

---

[1] See [1] for a good review of contemporary techniques in triaxial testing.

of the deformation are inevitable, as they are due to inherent instability of the material. At any rate, it should always be taken into account that measurements in triaxial tests become less reliable with increasing deformation.

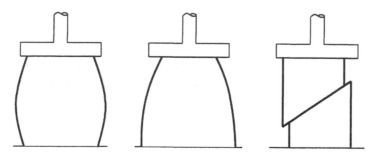

Figure 2.6: Inhomogeneous sample deformation modes

2.1.5 Behaviour of undrained samples

The behaviour of water-saturated undrained samples is completely different than the behaviour of drained ones. Measurement of the pore pressure u makes possible to determine the effective stress: $\sigma'_{ij} = \sigma_{ij} - u\delta_{ij}$. Herein the sign convention for pore pressure and for stress is the same: positive for compression. A complete saturation is important for successful tests. It is checked by the so-called B-test: The ratio $B := \Delta u / \Delta \sigma_2$ should equal 1 for a completely saturated sample.

The path of effective stresses during a triaxial test depends not only on the void ratio of the sample but also on the initial stress level, i.e. the σ'_2-value. The effective stress paths are represented either in a σ'_1-σ'_2-diagram (see Fig. 2.7) or as a diagram of $q := \sigma_1 - \sigma_2 = \sigma'_1 - \sigma'_2$ plotted over $p' := (\sigma'_1 + \sigma'_2 + \sigma'_3)/3$ (see Fig. 2.8). The corresponding pore pressure is shown in Fig. 2.9. The stress path A ends at the point R (see Fig. 2.8), which corresponds to the residual strength of the considered material. In general we can distinguish between the stress path patterns of types A, B and C (see Fig. 2.8). The type is determined by the initial state of the sample as represented in a e-σ'_2-diagram (see Fig. 2.10): Considering one and the same initial lateral stress σ'_2, a loose sample (A) will exhibit softening (i.e. decrease of stress deviator q) whereas a less loose sample (B) will initially exhibit softening and later hardening (i.e. increase of stress deviator q). A dense sample (C) will not exhibit any softening. Now the decision of whether a sample is to be considered as 'loose' or 'dense' in the above sense does not only depend on the void ratio e but also on the initial lateral stress σ'_2. The corresponding regions in a e-σ'_2-space are separated by the lines 1-1, 2-2 and 3-3, as shown in Fig. 2.10. A starting point above 1-1 leads to a stress path of type A, a starting point between 1-1 and 2-2 leads to type B, and a starting point between 2-2 and 3-3 leads to type C.

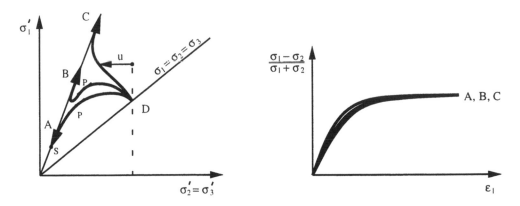

Figure 2.7: Stress paths and stress-strain curves at undrained triaxial tests

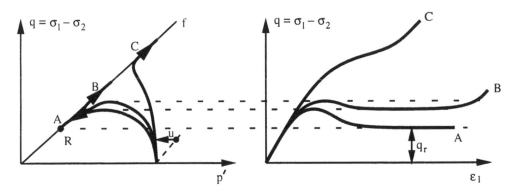

Figure 2.8: Stress paths and stress-strain curves at undrained triaxial tests, alternative representation

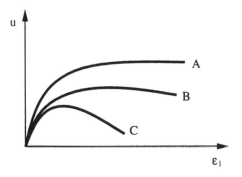

Figure 2.9: Development of pore pressure at undrained tests

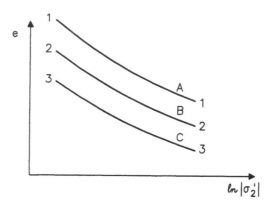

Figure 2.10: Regions in the e-σ'_2-space

2.1.6   The true triaxial test

The biaxial test according to HAMBLY's principle (see Fig. 2.11) makes possible to apply arbitrary rectilinear extensions, i.e. to change the sides of the rectangular cross section of a sample in plane deformation. This principle can also be applied in

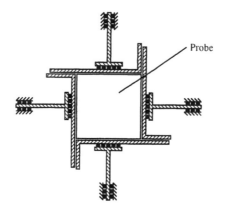

Figure 2.11: HAMBLY's principle for biaxial test

three dimensions. The corresponding apparatuses are highly complex and only a few prototypes exist in advanced research laboratories[2]. They make possible to explore a large variety of stress paths of so-called rectilinear extensions, i.e. motions without rotation of the principal stress and strain axes. Based on tests obtained with a true triaxial apparatus GOLDSCHEIDER formulated a principle according to which [14]

---

[2]e.g. in Cambridge, Karlsruhe, Grenoble

Proportional (i.e. straight) strain paths starting from a (nearly) stress free state are connected with proportional stress paths. If the initial state is not stress free, then the obtained stress path approaches asymptotically the path starting from the stress free state (see Fig. 2.12).

This theorem has far-reaching consequences.

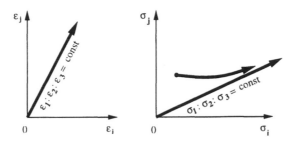

Figure 2.12: Stress and strain paths referring to GOLDSCHEIDER's principle

2.2 Oedometric test

The oedometric test is to simulate one-dimensional compression. In most cases the soil sample is loaded in vertical direction, whereas rigid side walls hinder any lateral expansion (see Fig. 2.13). The obtained stress-strain curves are non-linear as shown

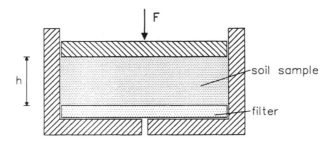

Figure 2.13: Principal array of an oedometer

in Fig. 2.14. More precisely, it comes out that the incremental (or tangential) stiffness $E_s := d\sigma_1/d\varepsilon_1$ increases (nearly linear) with σ_1. Expressing the compression by the change of the void ratio e, the behaviour of soils can be approximately described by the relation

$$d\sigma_1 = -C_c^{-1}\sigma_1 de \quad , \tag{2.1}$$

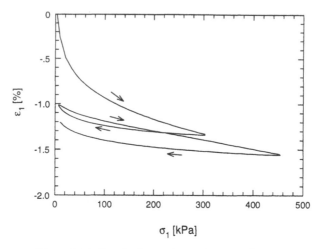

Figure 2.14: Result of an oedometric test with sand

or equivalently

$$e = e_0 - C_c \ln \frac{\sigma_1}{\sigma_0} \quad . \tag{2.2}$$

Equ. (2.1) leads to

$$E_s = \frac{1+e}{C_c}\sigma_1 \quad .$$

More precisely, investigations show that E_s is underproportional to σ_1, i.e. $E_s \propto \sigma_1^\alpha$ with $0 < \alpha < 1$. For unloading (so-called swelling) the C_c-values should be replaced by C_s with $C_s \approx 10C_c$. If we measure the lateral (horizontal) stress σ_2 during oedometric loading and subsequent unloading we observe that σ_2 increases proportionally to σ_1, i.e. $\sigma_2 = K_0\sigma_1$ with K_0 being the so-called earth pressure coefficient at rest. Upon unloading σ_2 reduces much less pronounced than σ_1 (see Fig. 2.15), i.e. an increased horizontal stress remains within the sample. This stress can be reduced by vibrations.

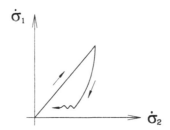

Figure 2.15: Stress path at oedometric loading and unloading

Chapter 3

Fundamentals of Continuum Mechanics

3.1 Deformation

A motion consists of translation, rotation and deformation. A material point with the material (or initial or LAGRANGE) coordinates $X_\alpha (\alpha = 1, 2, 3)$ moves into a position with the spatial (or EULER) coordinates $x_i (i = 1, 2, 3)$. Thus, the motion is described by the function $\mathbf{x} = \chi(\mathbf{X}, t)$. Using a less exact notation we can write $\mathbf{x} = \mathbf{x}(\mathbf{X}, t)$. The deformation gradient is defined as

$$\mathbf{F} = F_{i\alpha} = x_{i,\alpha} = \frac{\partial x_i}{\partial X_\alpha} = \frac{\partial \mathbf{x}}{\partial \mathbf{X}}$$

and can be decomposed into

$$\mathbf{F} = \mathbf{RU} = \mathbf{VR} \quad .$$

Herein the rotation $\mathbf{R}$ is a properly orthogonal tensor, i.e. $\mathbf{R}^T = \mathbf{R}^{-1}$, $\det \mathbf{R} = 1$. $\mathbf{U}$ and $\mathbf{V}$ are the right and left stretch tensors, respectively. The right CAUCHY-GREEN deformation tensor is defined as

$$\mathbf{C} := \mathbf{U}^2 = \mathbf{F}^T \mathbf{F} \quad ,$$

and the left CAUCHY-GREEN deformation tensor is defined as

$$\mathbf{B} := \mathbf{V}^2 = \mathbf{F}\mathbf{F}^T = \mathbf{R}\mathbf{C}\mathbf{R}^T \quad .$$

3.2 Stretching

EULER's stretching tensor $\mathbf{D}$ is obtained as the symmetric part of the velocity gradient $\mathbf{L} = \operatorname{grad} \mathbf{v} = v_{i,j} = \dot{x}_{i,j}$. Thus we have

$$\mathbf{D} = D_{ij} = \frac{1}{2}(v_{i,j} + v_{j,i}) = \frac{1}{2}(\dot{x}_{i,j} + \dot{x}_{j,i}) = \dot{x}_{(i,j)}. \tag{3.1}$$

We can also write:

$$\mathbf{D}(t) = \dot{\mathbf{U}}_{(t)}(t) = \frac{1}{2}\mathbf{R}(\dot{\mathbf{U}}\mathbf{U}^{-1} + \mathbf{U}^{-1}\dot{\mathbf{U}})\mathbf{R}^T \quad , \tag{3.2}$$

where the notation $\mathbf{U}_{(t)}$ denotes that the reference configuration of $\mathbf{U}$ is taken at the time t. One can show ([68], §82), that the time rate of the square of the differential arc length reads:

$$\overline{\dot{\mathrm{d}s^2}} = 2D_{ij}\mathrm{d}x_i\mathrm{d}x_j. \tag{3.3}$$

A vanishing small line of the length l in the direction $\mathbf{n}$ has the following stretching rate:

$$\lim_{l \to 0} \frac{\dot{l}}{l} = D_{ij}n_in_j. \tag{3.4}$$

CAUCHY's spin tensor is obtained as the antimetric part of the velocity gradient:

$$\mathbf{W} = W_{ij} = \frac{1}{2}(v_{i,j} - v_{j,i}) = \frac{1}{2}(\dot{x}_{i,j} - \dot{x}_{j,i}) = \dot{x}_{[i,j]}.$$

A motion with $W_{ij} = 0$ is called irrotational. If the normal velocities on the boundary of a region are prescribed, then irrotational motions minimize the kinetic energy within this region (KELVIN).

Note that the reference configuration for the tensor $\mathbf{D}$ is the actual one. $\mathbf{D}$ should not be confused with the time rate of any finite strain [21]. Following [68], §95 we can write:

$$\dot{C}_{\alpha\beta} = 2\dot{E}_{\alpha\beta} = 2D_{ij}x_{i,\alpha}x_{j,\beta}. \tag{3.5}$$

Neither should $\mathbf{W}$ be confused with the time rate of $\mathbf{R}$ ($\mathbf{W} \neq \dot{\mathbf{R}}$). The equality is only valid if the reference configuration is identical with the actual one, i.e.

$$\mathbf{W}(t) = \dot{\mathbf{R}}_{(t)}(t) \quad .$$

3.3   Simple shear

In soil mechanics the notion 'simple shear' encomprises also shear with volume change, i.e. dilatant shear. During this shear the material point (X_1, X_2) obtains the position (x_1, x_2):

$$x_1(t) = X_1 + X_2 \cdot f(t) \quad , \tag{3.6}$$

$$x_2(t) = X_2 + X_2 \cdot g(t) \quad . \tag{3.7}$$

We require that for $t = 0$ the material coordinates coincide with the spatial ones, hence $f(0) = g(0) = 0$.

From equ. (3.7) in follows $x_2 = (1 + g)X_2$. From (3.6) and (3.7) we obtain the deformation gradient

$$\mathbf{F} = \frac{\partial \mathbf{x}}{\partial \mathbf{X}} = \begin{pmatrix} 1 & f \\ 0 & (1+g) \end{pmatrix} \quad .$$

The velocity gradient $\mathbf{L} = \text{grad}\,\dot{\mathbf{x}} = \dfrac{\partial \dot{x}_i}{\partial x_j}$ can be obtained from

$$\dot{x} = \begin{pmatrix} X_2 \dot{f} \\ X_2 \dot{g} \end{pmatrix} = \frac{1}{1+g} \begin{pmatrix} x_2 \dot{f} \\ x_2 \dot{g} \end{pmatrix} \quad . \tag{3.8}$$

Its symmetric and antimetric parts read

$$\begin{aligned} \mathbf{D} &= \frac{1}{2(1+g)} \begin{pmatrix} 0 & \dot{f} \\ \dot{f} & 2\dot{g} \end{pmatrix} , \\ \mathbf{W} &= \frac{1}{2(1+g)} \begin{pmatrix} 0 & \dot{f} \\ -\dot{f} & 0 \end{pmatrix} . \end{aligned} \tag{3.9}$$

or, in three dimensions:

$$\begin{aligned} \mathbf{D} &= \frac{1}{2(1+g)} \begin{pmatrix} 0 & \dot{f} & 0 \\ \dot{f} & 2\dot{g} & 0 \\ 0 & 0 & 0 \end{pmatrix} , \\ \mathbf{W} &= \frac{1}{2(1+g)} \begin{pmatrix} 0 & \dot{f} & 0 \\ -\dot{f} & 0 & 0 \\ 0 & 0 & 0 \end{pmatrix} . \end{aligned} \tag{3.10}$$

The velocity field for the dilatant shear of Fig. 3.2 can be represented as

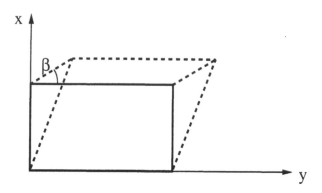

Figure 3.1: Simple shear

$$\mathbf{v} = \frac{1}{d}(\mathbf{n} \cdot \mathbf{x})\mathbf{v}_0 \quad .$$

With $\mathbf{v}_0(\mathbf{n} \cdot \mathbf{x}) \equiv (\mathbf{v}_0 \otimes \mathbf{n})\mathbf{x}$ it follows

$$\mathbf{v} = \frac{1}{d}(\mathbf{v}_0 \otimes \mathbf{n})\mathbf{x} \quad .$$

Thus, the velocity gradient $\mathbf{L} = \dfrac{\partial \mathbf{v}}{\partial \mathbf{x}}$ can be represented as

$$\mathbf{L} = \frac{1}{d}\mathbf{v}_0 \otimes \mathbf{n} \quad .$$

3.4   Cauchy stress

Cutting a body reveals the internal forces acting within it. Let us consider a particular point of the cutting surface with the unit normal $\mathbf{n}$ and the stress vector (i.e. areal density of force) $\mathbf{t}$. Both vectors are connected by the linear transformation $\mathbf{T}$:

$$\mathbf{t} = \mathbf{Tn}.$$

$\mathbf{T}$ is the CAUCHY stress tensor. By lack of couple stresses the stress tensor $\mathbf{T}$ is symmetric. $\mathbf{T}$ can be decomposed in a deviatoric ($\mathbf{T}^*$) and a hydrostatic part ($\frac{1}{3}\mathrm{tr}\mathbf{T1}$):

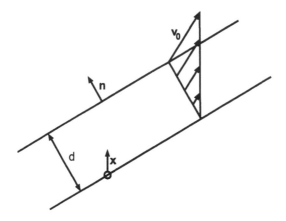

Figure 3.2: Simple shear within a shear band of thickness d

$$\mathbf{T} = \mathbf{T}^* + \frac{1}{3}(\mathrm{tr}\mathbf{T})\mathbf{1}$$

where $\mathrm{tr}\,\mathbf{T}$ denotes the sum $T_1 + T_2 + T_3$.

The components of $\mathbf{T}$ are denoted by T_{ij}:

$$\begin{pmatrix} T_{11} & T_{12} & T_{13} \\ T_{21} & T_{22} & T_{23} \\ T_{31} & T_{32} & T_{33} \end{pmatrix} .$$

T_{ij} is the component of $\mathbf{T}$ in i-direction acting upon a surface with unit normal $\mathbf{e}_j$. T_{11}, T_{22}, T_{33} are normal stresses (negative at compression), $T_{12} = T_{21}$, $T_{31} = T_{13}$, $T_{23} = T_{32}$ are shear stresses. The deviatoric part of $\mathbf{T}$ has the components

$$T_{11}^* = \frac{1}{3}(2T_{11} - T_{22} - T_{33})$$

$$T_{22}^* = \frac{1}{3}(2T_{22} - T_{33} - T_{11})$$

$$T_{33}^* = \frac{1}{3}(2T_{33} - T_{11} - T_{22}) .$$

The mixed components are unchanged, i.e. $T_{12}^* = T_{12}$ etc.

In traditional soil mechanics compressive principal stresses are taken as positive. To avoid confusion we introduce $\sigma_{ij} := -T_{ij}$ as the CAUCHY-stress with the sign convention of traditional soil mechanics.

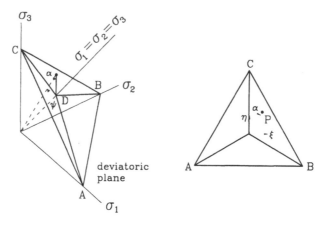

Figure 3.3: Deviatoric plane

Some results in soil mechanics are represented in the tree-dimensional space of principal stresses T_1, T_2, T_3 (or $\sigma_1, \sigma_2, \sigma_3$). This is only meaningful for rectilinear extensions, i.e. for motions without rotation of principal stresses. Hydrostatic stress states lie on the principal space diagonal, whereas the equation $\mathrm{tr}\,\mathbf{T} = \mathrm{const}$ defines so-called deviatoric planes. The position of a stress point on a deviatoric plane can be determined by the angles α and ψ (see 3.3). α is defined by the equation

$$\cos(3\alpha) = \sqrt{6}\,\frac{\mathrm{tr}\left(\mathbf{T}^{*3}\right)}{(\mathrm{tr}\,\mathbf{T}^{*2})^{3/2}} \quad,$$

i.e. α depends only on the deviatoric part $\mathbf{T}^*$. ψ represents the angular deviation from the principal space diagonal.

$$\cot\psi = \frac{\mathrm{tr}\,\mathbf{T}}{\sqrt{3\mathrm{tr}\left(\mathbf{T}^{*2}\right)}} \quad.$$

Points in a deviatoric plane can also be characterized with the Cartesian co-ordinates ξ and η:

$$
\begin{aligned}
\xi &= \frac{1}{\sqrt{2}}(T_2 - T_1) \\
\eta &= \frac{1}{\sqrt{6}}(2T_3 - T_1 - T_2) = \frac{3}{\sqrt{6}}T_3^* \quad.
\end{aligned}
\tag{3.11}
$$

3.5   Change in observer

Let $\mathbf{x}(\mathbf{X})$ be a motion. A so-called equivalent motion $\mathbf{x}^*$ is obtained from $\mathbf{x}$ by a *change in observer* if

$$\mathbf{x}^*(\mathbf{X}, t) = \mathbf{q}(t) + \mathbf{Q}(t)[\mathbf{x}(\mathbf{X}, t) - \mathbf{o}] \tag{3.12}$$

$\mathbf{x}$ and $\mathbf{x}^*$ differ by a rigid body motion, which consists of the translation $\mathbf{q}(t)$ and the rotation $\mathbf{Q}(t)$. With $\mathbf{F} = \partial\mathbf{x}/\partial\mathbf{X}$ and $\mathbf{F}^* = \partial\mathbf{x}^*/\partial\mathbf{X}$ it follows from (3.12)

$$\mathbf{F}^*(\mathbf{X}, t) = \mathbf{Q}(t)\mathbf{F}(\mathbf{X}, t) \quad . \tag{3.13}$$

With $\mathbf{F} = \mathbf{RU} = \mathbf{VR}$ and $\mathbf{F}^* = \mathbf{R}^*\mathbf{U}^* = \mathbf{V}^*\mathbf{R}^*$ follows from (3.13)

$$\mathbf{F}^* = \mathbf{R}^*\mathbf{U}^* = \mathbf{QF} = \mathbf{QRU} \quad . \tag{3.14}$$

As $\mathbf{QR}$ is a rotation, we obtain from (3.14) :

$$\mathbf{R}^* = \mathbf{QR}, \quad \mathbf{U}^* = \mathbf{U} \quad . \tag{3.15}$$

Introducing (3.15) into $\mathbf{V}^* = \mathbf{R}^*\mathbf{U}^*\mathbf{R}^{*T}$ we obtain:

$$\mathbf{V}^* = \mathbf{QRUR}^T\mathbf{Q}^T \quad . \tag{3.16}$$

Herefrom and with $\mathbf{V} = \mathbf{RUR}^T$ it follows

$$\mathbf{V}^* = \mathbf{QVQ}^T \quad . \tag{3.17}$$

Thus, the CAUCHY-GREEN tensors $\mathbf{C} = \mathbf{U}^2$ and $\mathbf{B} = \mathbf{V}^2$ are transformed as follows:

$$\mathbf{C}^* = \mathbf{C}, \quad \mathbf{B}^* = \mathbf{QBQ}^T \quad . \tag{3.18}$$

(3.18b) follows from $\mathbf{V}^{*2} = \mathbf{V}^{*T}\mathbf{V}^* = \mathbf{QV}^T\mathbf{Q}^T\mathbf{QVQ}^T = \mathbf{QV}^2\mathbf{Q}^T$.

Now we differentiate (3.12) with respect to t:

$$\dot{\mathbf{x}}^*(\mathbf{X}, t) = \dot{\mathbf{q}}(t) + \mathbf{Q}(t)\dot{\mathbf{x}}(\mathbf{X}, t) + \dot{\mathbf{Q}}(t)(\mathbf{x} - \mathbf{o})$$

With $\mathbf{v}^*(\mathbf{x}^*, t) = \dot{\mathbf{x}}^*(\mathbf{X}, t)$ and $\mathbf{v}(\mathbf{x}, t) = \dot{\mathbf{x}}(\mathbf{X}, t)$ we obtain the transformation rule for $\mathbf{v}$:

$$\mathbf{v}^*(\mathbf{x}^*, t) = \dot{\mathbf{q}}(t) + \mathbf{Q}(t)\mathbf{v}(\mathbf{x}, t) + \dot{\mathbf{Q}}(t)(\mathbf{x} - \mathbf{o}) \tag{3.19}$$

Thus we can obtain the transformation rule for the velocity gradient, i.e. a relation between $\mathbf{L} = \operatorname{grad} \mathbf{v}$ and $\mathbf{L}^* = \operatorname{grad} \mathbf{v}^*$. Differentiating (3.19) with respect to $\mathbf{x}$

$$\underbrace{\frac{\partial \mathbf{v}^*}{\partial \mathbf{x}^*}}_{\mathbf{L}^*} \underbrace{\frac{\partial \mathbf{x}^*}{\partial \mathbf{x}}}_{\mathbf{Q}(t)} = \mathbf{Q}(t)\mathbf{L}(\mathbf{x}, t) + \dot{\mathbf{Q}}(t) \tag{3.20}$$

and multiplying (3.20) with $\mathbf{Q}^T$ we obtain

$$\mathbf{L}^* = \mathbf{Q}\mathbf{L}\mathbf{Q}^T + \dot{\mathbf{Q}}\mathbf{Q}^T \quad . \tag{3.21}$$

If we differentiate $\mathbf{Q}\mathbf{Q}^T = 1$ with respect to t, we obtain $\dot{\mathbf{Q}}\mathbf{Q}^T = -(\dot{\mathbf{Q}}\mathbf{Q}^T)^T$. Herefrom it follows that $\dot{\mathbf{Q}}\mathbf{Q}^T$ is an antimetric tensor. $\mathbf{D}$ is defined as the symmetric part of $\mathbf{L}$:

$$\mathbf{D} = \frac{1}{2}(\mathbf{L} + \mathbf{L}^T), \quad \mathbf{D}^* = \frac{1}{2}(\mathbf{L}^* + \mathbf{L}^{*T}),$$

thus we obtain from (3.21):

$$\mathbf{D}^* = \frac{1}{2}(\mathbf{Q}\mathbf{L}\mathbf{Q}^T + \mathbf{Q}\mathbf{L}^T\mathbf{Q}^T) = \mathbf{Q}\frac{1}{2}(\mathbf{L} + \mathbf{L}^T)\mathbf{Q}^T = \mathbf{Q}\mathbf{D}\mathbf{Q}^T \quad . \tag{3.22}$$

The spin $\mathbf{W}$ is transformed as follows:

$$\mathbf{W}^* = \mathbf{Q}\mathbf{W}\mathbf{Q}^T + \dot{\mathbf{Q}}\mathbf{Q}^T \quad . \tag{3.23}$$

Let us now see how to transform the stress tensor $\mathbf{T}$ at change of the reference configuration. With $\mathbf{n}^* = \mathbf{Q}\mathbf{n}$ and $\mathbf{t}^* = \mathbf{Q}\mathbf{t}$ we obtain from $\mathbf{t}^* = \mathbf{T}^*\mathbf{n}^*$:

$$\mathbf{Qt} = \mathbf{T^*Qn}$$

or

$$\mathbf{t} = \mathbf{Q}^T\mathbf{T^*Qn} = \mathbf{Tn} \quad .$$

It then follows

$$\mathbf{T^*} = \mathbf{QTQ}^T \quad . \tag{3.24}$$

3.6 Objectivity, objective time rates

The material behaviour is called *independent of the observer* if the stress is transformed according to (3.24) . All tensors transformed according to (3.24) are called *independent of the observer* or *indifferent*.

A co-rotated observer registers the stress $\mathbf{T^*} = \mathbf{QTQ}^T$. If the observer is at rest and the considered material is rotated by $\mathbf{R}$ (with $\mathbf{Q} = \mathbf{R}^T$), then the observer registers the stress $\mathbf{T^*} = \mathbf{R}^T\mathbf{TR}$. Thus, $\mathbf{T^*}$ is the co-rotated stress. The observer registers the following time rate of $\mathbf{T^*}$.

$$\dot{\mathbf{T}}^* = \dot{\mathbf{R}}^T\mathbf{TR} + \mathbf{R}^T\dot{\mathbf{T}}\mathbf{R} + \mathbf{R}^T\mathbf{T}\dot{\mathbf{R}} \quad .$$

Now we choose the actual configuration as our reference configuration. Then we have $\mathbf{R} = \mathbf{R}^T = 1$ and $\dot{\mathbf{R}} = -\dot{\mathbf{R}}^T = \mathbf{W}$, and we denote $\dot{\mathbf{T}}^*$ as $\overset{\circ}{\mathbf{T}}$:

$$\overset{\circ}{\mathbf{T}} = \dot{\mathbf{T}} - \mathbf{WT} + \mathbf{TW} \quad .$$

$\overset{\circ}{\mathbf{T}}$ is the co-rotational or ZAREMBA[1] stress rate. $\overset{\circ}{\mathbf{T}}$ is the stress change that results solely from the deformation of the considered material, whereas any apparent parts (due to rotations of the observer or of the reference frame) are removed.

---

[1] Often attributed to JAUMANN

The *principle of material frame-indifference*, shortly called *objectivity*, requires that a constitutive equation determines the stress $\mathbf{T}$ in such a way, that an equivalent motion leads to $\mathbf{T}^*$, whereas $\mathbf{T}^*$ and $\mathbf{T}$ are related by $\mathbf{T}^* = \mathbf{Q}\mathbf{T}\mathbf{Q}^T$.

In general, derivation of vectorial and tensorial quantities with respect to time imposes problems if they refer to a material (also called 'concomitant' or 'convected') vector basis: A mere change of the reference frame gives rise to a non-vanishing time derivative of the considered quantity. This time derivative is apparent (or non-objective), as it depends on the equivalent motion (or change of the reference frame), which is arbitrary. There are several ways how to introduce objective time rates, i.e. time rates that are free of apparent terms[2]. Considering the CAUCHY stress $\mathbf{T}$ (which in itself is an objective quantity) the following objective time rates have been introduced

$$
\begin{aligned}
&\text{ZAREMBA or JAUMANN:} & \overset{\circ}{\mathbf{T}} &:= \dot{\mathbf{T}} - \mathbf{W}\mathbf{T} + \mathbf{T}\mathbf{W} \\
&\text{LIE or OLDROYD:} & \mathcal{L}\mathbf{T} &:= \dot{\mathbf{T}} - \mathbf{L}\mathbf{T} - \mathbf{T}\mathbf{L}^T \\
&\text{convected stress rate:} & \overset{\triangle}{\mathbf{T}} &:= \dot{\mathbf{T}} + \mathbf{L}^T\mathbf{T} + \mathbf{T}\mathbf{L} \\
&\text{GREEN-MCINNIS-NAGHDI}\ [3] & \overset{\square}{\mathbf{T}} &:= \dot{\mathbf{T}} - \boldsymbol{\Omega}\mathbf{T} + \mathbf{T}\boldsymbol{\Omega}
\end{aligned}
$$

with[4] $\boldsymbol{\Omega} := \dot{\mathbf{R}}\mathbf{R}^T = \mathbf{W} - \tfrac{1}{2}\mathbf{R}(\dot{\mathbf{U}}\mathbf{U}^{-1} - \mathbf{U}^{-1}\dot{\mathbf{U}})\mathbf{R}^T$.

The differences between individual objective stress rates are themselves objective quantities, e.g.

$$
\overset{\circ}{\mathbf{T}} - \mathcal{L}\mathbf{T} = \mathbf{D}\mathbf{T} + \mathbf{T}\mathbf{D}\ \ .
$$

We see thus that we can obtain an infinite number of objective stress rates by adding to $\overset{\circ}{\mathbf{T}}$ objective quantities such as $\mathbf{D}\mathbf{T} + \mathbf{T}\mathbf{D}$ or $(\mathrm{tr}\,\mathbf{T})\mathbf{D}$. Such quantities are possible terms of constitutive equations of the rate type. Thus, a discussion on the proper objective stress rate is, virtually, a discussion on the proper constitutive law. This is why TRUESDELL and NOLL write

> Clearly the properties of a material are *independent of the choice of flux*[5] which, like the choice of a measure of strain, is *absolutely immaterial*. ... Thus we leave intentionally uncited the blossoming literature on invariant time fluxes subjected to various arbitrary requirements.

---

[2]see [67] sect. 36, [20] Chapter VII, [75] p. 45, [61] p. 254
[4]see [67], equ. (24.16)
[5]time rate is also called 'flux'

3.7 General constitutive equation

Following the *principle of determinism of the stress* and the *principle of local action* the stress within a non-polar material at a given time t, $\mathbf{T}(t)$, depends on the previous history of the motion χ of a neighbourhood of a material point $\mathbf{X}$. This history is represented as a functional

$$\mathbf{T}(t) = \mathcal{F}_t(\chi) \quad .$$

For an equivalent motion χ^* objectivity requires:

$$\mathbf{T}^*(t^*) = \mathcal{F}_{t^*}(\chi^*) \quad .$$

For so-called *simple materials* the stress depends only on the history of the deformation gradient:

$$\mathbf{T}(t) = \mathcal{G}(\mathbf{F}^t) = \mathcal{G}_{s=0}^{\infty}(\mathbf{F}(t - s))$$

whereas for so-called materials of the grade n also the n-th deformation gradient is important[6].

3.8 Principle of macrodeterminism

We consider two strain paths with identical initial and end points (see Fig. 3.4). The one path is smooth, whereas the other path results from the smooth one by superposition of small deviations. Let $\Delta\varepsilon$ be the maximum deviation between the two paths. These deviations may result e.g. from an automatic control obtained when trying to pursue the smooth path. The question is whether the corresponding maximum stress deviation $\Delta\sigma$ is large or not. More precisely, it is interesting to observe whether $\Delta\sigma \to 0$ implies $\Delta\varepsilon \to 0$ or not. The first case constitutes the so-called *principle of macrodeterminism* [28, 35]. By now there are not experimental results either to disprove or to corroborate this principle for real materials. Thus, it is still a (rather questionable) postulate than a principle. It is not fulfilled by hypoplastic equations.

---

[6]see footnote in [67] p. 63

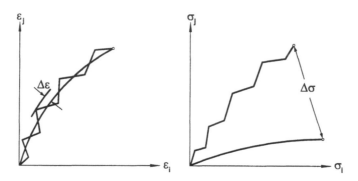

Figure 3.4: Smooth and zig-zag-paths

3.9   Internal constraints

An internal constraint is given if the possible motions of a material point are constrained by a requirement

$$\gamma(\mathbf{F}) = 0 \quad . \tag{3.25}$$

An internal constraint is a sort of constitutive equation and should obey the principle of objectivity. Thus from (3.25) it follows

$$\lambda(\mathbf{C}(\tau)) = 0, \qquad -\infty < \tau < \infty \quad , \tag{3.26}$$

where $\mathbf{C}$ is the right CAUCHY-GREEN tensor. Differentiation with respect to t yields:

$$\mathrm{tr}\left(\frac{\partial\lambda}{\partial\mathbf{C}}\dot{\mathbf{C}}\right) = 0 \quad .$$

With $\dot{\mathbf{C}} = 2\mathbf{F}^T\mathbf{D}\mathbf{F}$ we obtain herefrom

$$\mathrm{tr}\left[\underbrace{\mathbf{F}\frac{\partial\lambda}{\partial\mathbf{C}}\mathbf{F}^T}_{=:\mathbf{A}}\mathbf{D}\right] = 0 \quad . \tag{3.27}$$

The history of $\mathbf{F}$ determines the stress $\mathbf{T}$ only up to a part $\mathbf{N}$. $\mathbf{N}$ is needed to fulfil the internal constraint but it does not dissipate energy:

$$\mathrm{tr}\left(\mathbf{N}\mathbf{D}\right) = 0 \quad . \tag{3.28}$$

From (3.27) and (3.28) follows that $\mathbf{N}$ differs from $\mathbf{A}$ only by a scalar factor q:

$$\mathbf{N} = q\mathbf{F}\frac{\partial\lambda}{\partial\mathbf{C}}\mathbf{F}^T \quad .$$

The following special internal constraints are often encountered:

Incompressibility: From the requirement for isochoric motions, $\det \mathbf{F} = 1$, it follows

$$\lambda(\mathbf{C}) = \det \mathbf{C} - 1 = 0 \quad .$$

Hence, the stress within incompressible materials can only be determined as

$$\mathbf{T} + u\mathbf{1} \quad ,$$

where u is constitutively undeterminate.

Rigidity: The entire stress tensor is constitutively undeterminate.

Inextensibility: Inextensibility in the direction $\mathbf{e}$ is given by the constraint

$$\lambda(\mathbf{C}) = \mathbf{e} \cdot \mathbf{C}\mathbf{e} - 1 = 0 \quad .$$

It follows:

$$\mathbf{N} = q\mathbf{Fe} \otimes \mathbf{Fe} \quad .$$

3.10 Effective stress

We consider a water-saturated granular medium and we assume that both water and grains are incompressible. Furthermore, we consider undrained and, thus, isochoric (i.e. volume preserving) deformation. The assumed incompressibility imposes an internal constraint and, therefore, the stress σ_{ij} can be determined from the deformation only up to the constitutively undetermined hydrostatic stress part u:

$$\sigma_{ij} = \sigma'_{ij} + u\delta_{ij} \quad . \tag{3.29}$$

Herein, we consider compressive stresses and pressures as positive. The constitutively determined stress, σ'_{ij}, is called the *effective stress*. It is important to note that u acts not only within the porewater but also within the grains and also at the interfaces between the individual grains (see Fig. 3.5).

TERZAGHI's definition of the *principle of effective stress* states:

> The stresses in any point of a section through a mass of soil can be computed from the total principal stresses, σ_1, σ_2, σ_3, which act in this point. If the voids of the soil are filled with water under a stress u, the

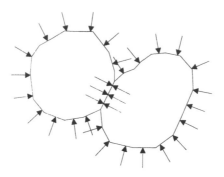

Figure 3.5: Interface stress between two grains caused by pore pressure

total principal stress consists of two parts. One part, u, acts in the water and in the solid in every direction with equal intensity. It is called the *neutral stress* (or the pore water pressure). The balance $\sigma_1' = \sigma_1 - u$, $\sigma_2' = \sigma_2 - u$ and $\sigma_3' = \sigma_3 - u$ represents an excess over the neutral stress u, and it has its seat exclusively in the solid phase of a soil. This fraction of the total principal stress will be called the *effective principal stress* .... A change in the neutral stress u produces practically no volume change and has practically no influence on the stress conditions of failure ... Porous materials (such as sand, clay and concrete) react to a change of u as if they were incompressible and as if their internal friction were equal to zero. All the measurable effects of a change of stress, such as compression, distortion and a change of shearing resistance are *exclusively* due to changes in the effective stresses σ_1', σ_2' and σ_3'. Hence every investigation of the stability of a saturated body of soil requires the knowledge of both the total and the neutral stresses.

3.11   Isotropy groups

There are particular deformations characterized by the deformation gradient $\mathbf{H}$, that cannot be detected by subsequent investigation of the material behaviour.

$$\mathcal{G}_{s=0}^{\infty}(\mathbf{F}^{(t)}(s)) = \mathcal{G}_{s=0}^{\infty}(\mathbf{F}^{(t)}(s)\mathbf{H}) \quad ,$$

All deformation gradients $\mathbf{H}$ with this property constitute the so-called isotropy (or symmetry) group of a material. An isotropy group is defined with reference to a particular configuration of the body. If an orthogonal tensor $\mathbf{Q}$ belongs to the isotropy group, we infer from objectivity:

$$\mathbf{Q}_0 \mathcal{G}_{s=0}^{\infty} (\mathbf{F}^{(t)}(s)) \mathbf{Q}_0^T = \mathcal{G}_{s=0}^{\infty} (\mathbf{Q}(s) \mathbf{F}^{(t)}(s)), \qquad \mathbf{Q}_0 := \mathbf{Q}(t=0) \quad,$$

Hence

$$\mathbf{Q} \mathcal{G}_{s=0}^{\infty} (\mathbf{F}^{(t)}(s)) \mathbf{Q}^T = \mathcal{G}_{s=0}^{\infty} (\mathbf{Q} \mathbf{F}^{(t)}(s) \mathbf{Q}^T) \quad.$$

A material is called isotropic if there is at least one undistorted state such that its isotropy group is the full orthogonal group. For isotropic materials we cannot detect rotations by means of mechanical tests.

3.12 Rate dependence

Rate-independent materials are defined as materials without an internal time scale. I.e., the rate of deformation is immaterial for the final stress. In other words, rate-independent materials are invariant with respect to changes of time scale. If we deform a rate-independent material twice as fast, then the stress rate will also be doubled. With respect to constitutive equations of the rate-type (i.e. constitutive equations of the type $\overset{\circ}{\mathbf{T}} = \mathbf{h}(\mathbf{T}, \mathbf{D})$), rate-independence means that the stress rate $\overset{\circ}{\mathbf{T}}$ is positively homogeneous of the first degree with respect to $\mathbf{D}$:

$$\mathbf{h}(\mathbf{T}, \lambda \mathbf{D}) = \lambda \mathbf{h}(\mathbf{T}, \mathbf{D}) \quad \text{for} \quad \lambda > 0 \quad.$$

Note that this homogeneity does by no means imply linearity (cf. the relation $y = |x|$, which is homogeneous in the above sense, but not linear). Soils are not exactly rate-independent. Clays are more pronouncedly rate dependent than sands, but also sands exhibit rate dependence [12]. However, for a first approximation we can consider soils as rate-independent materials.

Referring to rate dependence, we should distinguish related notions such as viscosity, creep, relaxation, time dependence. All these notions have in common that they imply one or more material parameters bearing the dimension of time. An explicit appearance of time t in the constitutive equation implies that there is no invariance with respect to the time coordinate. In other words, the material exhibits ageing.

In soil mechanics, rate dependence is mainly manifest as creep (cf. secondary consolidation) or increase of shear stresses by increasing the deformation rate. It resides in the grain skeleton and not in the free pore fluid (cf. consolidation according to TERZAGHI's theory).

Another phenomenon is relaxation, i.e. the decrease of stress with time when the deformation remains constant. The measurement of relaxation is very difficult as the measurement of forces by means of load cells requires a minute deformation of the load cell and, consequently, also of the sample such that the condition $\varepsilon = $ const is not perfectly fulfilled. The often assumed correspondence between relaxation and creep is directly applicable to viscoelastic materials only. For the general case, it is a very difficult task to incorporate relaxation in a constitutive relation. Perhaps this task is also of minor importance since the stress drop due to relaxation is quickly recovered as soon as the deformation is resumed.

Chapter 4

Hypoplasticity

4.1 Rate equations

A constitutive equation is expected to represent stress due to a strain (or deformation) history starting from some specified reference state. If we represent stress as a *function* of strain, this automatically means that the stress does not depend on the deformation *history*. This special case is called (by definition) elastic behaviour. Soil is not elastic, so we have to find another type of relation. How can we represent strain history? Some researchers introduced integral transformations using appropriate kernels. This approach is not useful for soils. A general way to introduce history (or path) dependence in physics is to use non-integrable differential forms (or PFAFFean forms), i.e. to represent y by the differential equation

$$dy = a_1 dx_1 + a_2 dx_2 + \ldots a_n dx_n.$$

This equation connects increments $dx_1, dx_2, \ldots$ with dy (or $dy_1, dy_2, \ldots$, if y is a vector) in such a way that there is no closed-form representation of $y(x)$. I.e. the relation (which is called incremental, as it relates increments) $dy = f(dx_i)$ is not integrable. This is the way we proceed in soil mechanics when we represent the stress increment as a non-integrable function of the strain increment:

$$d\sigma = f(d\epsilon) \quad .$$

This approach is common to the theories of plasticity and hypoplasticity.

Now we can divide all increments by dt and obtain time rates:

$$\dot\sigma = d\sigma/dt \quad , \quad \dot\epsilon = d\epsilon/dt \quad , \qquad \text{etc.}$$

Thus, an equation between increments is also representable as an equation between rates, as long as we refer to so-called rate-independent materials. An equation of the form $\dot\sigma = f(\dot\epsilon)$ is called a rate-equation. It does not imply the existence of an equation $\sigma = g(\epsilon)$.

In tensor notation a constitutive equation of the rate type has the form $\overset{\circ}{\mathbf{T}} = \mathbf{f}(\mathbf{D})$. It is often reasonable to include $\mathbf{T}$ in the list of arguments, i.e. to write $\overset{\circ}{\mathbf{T}} = \mathbf{f}(\mathbf{T}, \mathbf{D})$.

Note that, strictly speaking, $\mathbf{D}$ is not the time rate of any strain measure, and also $\overset{\circ}{\mathbf{T}} \neq \dot{\mathbf{T}}$. However, for the special case of rectilinear extensions ($\mathbf{W} = \mathbf{0}$) we have $\overset{\circ}{\mathbf{T}} = \dot{\mathbf{T}}$, and $\mathbf{D}$ is the time rate of logarithmic strain ϵ_{ij}.

4.2 Incremental non-linearity

$d\sigma/d\epsilon = \dot{\sigma}/\dot{\epsilon}$ represents the incremental stiffness of the material considered (see Fig. 4.1). Since for anelastic (plastic) materials $|d\sigma|$ is much larger at unloading than

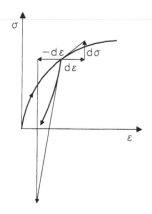

Figure 4.1: Different stiffness at loading and unloading

at loading (i.e. the stiffness is much larger at unloading than at loading), we infer that for such a material the function $d\sigma = f(d\epsilon)$ or $\dot{\sigma} = f(\dot{\epsilon})$ must be nonlinear in $\dot{\epsilon}$ (or $d\epsilon$). This non-linearity remains, no matter how small $d\epsilon$ is. Therefore it is called 'non-linearity in the small' or 'incremental non-linearity'. Note that incremental non-linearity has nothing to do with the curved form of the stress-strain curve for loading. This curve can be, of course, linearized for small $|d\epsilon|$, a fact which led many people to believe that in physics every relation can be linearized 'in the small'. Thus, all elastoplastic and hypoplastic relations are incrementally non-linear.

Incremental non-linearity is the seat of the hysteresis loop exhibited by stress-strain curves at cyclic stress. It also implies that constitutive equations of the form $\overset{\circ}{\mathbf{T}} = \mathbf{f}(\mathbf{T}, \mathbf{D})$ are non-linear in $\mathbf{D}$ and, also, non-differentiable at $\mathbf{D} = \mathbf{0}$. This fact imposes many mathematical difficulties.

4.3 Homogeneity in stress

Assume that the relation $\overset{\circ}{\mathbf{T}} = \mathbf{h}(\mathbf{T}, \mathbf{D})$ is homogeneous in $\mathbf{T}$, i.e.

$$\mathbf{h}(\lambda\mathbf{T}, \mathbf{D}) = \lambda^n \mathbf{h}(\mathbf{T}, \mathbf{D}) \quad .$$

Let us investigate the consequences of this assumption. Consider a stress state $\mathbf{T}_1$. We now determine the stretching in such a way that $\overset{\circ}{\mathbf{T}} = \mathbf{h}(\mathbf{T}_1, \mathbf{D}_1) = \lambda\mathbf{T}_1$. If we then continuously apply $\mathbf{D}_1$, then we shall obtain a stress path which is a straight line passing through the origin of stress space. This follows from our assumption, because

$$\overset{\circ}{\mathbf{T}}(t + dt) = \mathbf{h}(\mathbf{T}_1 + \lambda\mathbf{T}_1 dt, \mathbf{D}_1) = (1 + \lambda dt)^n \mathbf{h}(\mathbf{T}_1, \mathbf{D}_1) = (1 + \lambda dt)^n \overset{\circ}{\mathbf{T}}(t) \quad .$$

In other words, our assumption of homogeneity in $\mathbf{T}$ implies that proportional strain paths (i.e. paths with $\mathbf{D} = $ const) are connected with proportional stress paths (i.e. straight stress paths passing through the origin of the stress space) and conforms, thus, with GOLDSCHEIDER's principle (see section 2.1.6).

Note that proportional stress paths must be limited within a fan, because there are also inaccessible (infeasible) stress states (see Fig. 4.2). E.g., a stress state with

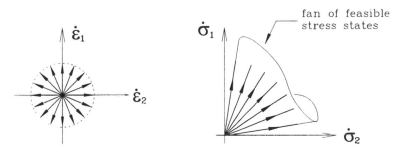

Figure 4.2: Strain path directions (left) and corresponding stress paths (right) obtained, each, by the constant application of the aforementioned strain directions and starting from the stress free state. Schematic representation showing only the components in 1- and 2-directions

tensile principal stresses is not feasible for cohesionless granulates. Referring to Fig. 2.12 it is interesting to note that if we apply the proportional strain path shown in its left part, which starts *not* from the stress-free state, we obtain the curved stress path shown in this figure. Let us now consider the degree of homogeneity. Knowing that $d\sigma/d\epsilon = \dot{\sigma}/\dot{\epsilon}$ or $\overset{\cdot}{\mathbf{T}}/\mathbf{D}$ is the stiffness, we infer that $(\overset{\cdot}{\mathbf{T}}/\mathbf{D})|_{\lambda\mathbf{T}} = \lambda^n (\overset{\cdot}{\mathbf{T}}/\mathbf{D})|_{\mathbf{T}}$. In other words, if we increase the stress by a factor λ, the stiffness is increased by the factor λ^n. Experimentalists in soil mechanics often remark that *normalized* stress-strain curves coincide (this is in particular the case with normally consolidated clays). The consequence is $n = 1$. Setting $n = 1$ would imply that the friction angle

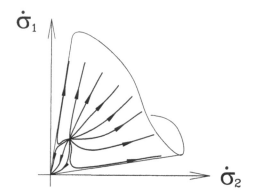

Figure 4.3: Stress paths obtained with proportional strain paths starting not from the stress free state

is invariant with respect to the stress level. This is an acceptable approximation to start with. If, however, the changes of stress level at a given void ratio are considerable, then the corresponding variation of friction (and dilatancy) angles may not be neglected. Note that for the case $n = 1$ all material constants must be dimensionless.

4.4   Hypoelasticity

TRUESDELL [66] has introduced constitutive relations of the form $\overset{\circ}{\mathbf{T}} = \mathbf{h}(\mathbf{T}, \mathbf{D})$. He required that the function $\mathbf{h}()$ be linear in $\mathbf{T}$ and in $\mathbf{D}$ and introduced the name hypoelasticity for such relations. Hypoelastic constitutive equations may produce curved stress-strain curves, and in some cases these stress-strain curves reach a horizontal plateau and can thus model yielding. However, the imposed incremental linearity implies equal stiffness for loading and unloading and thus renders hypoelastic relations inappropriate to describe anelastic (plastic) materials. Despite this, some hypoelastic relations have been launched in soil mechanics (e.g. by DAVIS and MULLENGER [9]). In order to avoid equal stiffness at loading and unloading, they are (in most cases tacitly) endowed with additional stress-strain relations holding for (appropriately defined) unloading. Strictly speaking, such relations (regarded as a whole) are not linear any more, i.e. they are not hypoelastic.

4.5   Elastoplasticity

The problem of describing different stiffness at loading and unloading can be treated by introducing at least two different linear relations between $\dot{\sigma}$ and $\dot{\varepsilon}$, of which one holds for loading and one for unloading. This is the approach of the theory of elastoplasticity. It requires a series of precautions. First, it has to be defined what should be considered as loading and what as unloading. This is accomplished by introducing the so-called yield surface, a surface in the stress space. Only such stress increments which start from this surface and point outwards are considered as loading, the remaining being unloading stress increments. The common theories of elastoplasticity require that the behaviour is elastic inside the yield surface, an assumption which is not realistic for soils. Another precaution refers to the expectation that in the transition between loading and unloading the response must be continuous. This is accomplished by the so-called consistency condition. Another point of concern in elastoplasticity is how the yield surface changes its shape and position with loading. It is typical for the theory of elastoplasticity to consider a decomposition of strain into elastic and plastic strain parts, which cannot be in fact distinguished in experiments. The stress-strain relation for loading is determined by the so-called flow rule, which states that the increment (or rate) of the plastic strain is always normal to a so-called plastic potential surface. A special case arises if the plastic potential surface is set equal to the yield surface. This special case is called the normality condition. A set of very useful theorems to estimate collapse loads has been formulated for materials obeying the normality condition. However, normality is not realistic for frictional soils, as it would imply that the dilatancy angle is equal to the friction angle, which is not the case. After all, we can summarize: Elastoplastic constitutive laws consist of two or more linear relations between $d\varepsilon$ and $d\sigma$. As a whole, they are incrementally non-linear.

The concepts of elastoplasticity are expressed mathematically as follows: Starting from the decomposition of strain into elastic and plastic parts,

$$\varepsilon_{ij} = \varepsilon_{ij}^e + \varepsilon_{ij}^p \quad ,$$

the yield function $f(\sigma_{ij}, \varepsilon_{ij}^p)$ is introduced such that the equation $f = 0$ defines the yield surface. If f does not depend on ε_{ij}^p we have the special case of the so-called ideal plasticity, whereas the (whatsoever defined) dependence of f on ε_{ij}^p is called hardening. By means of the yield function we can define *loading* by the conditions

$$f = 0 \quad \text{and} \quad \frac{\partial f}{\partial \sigma_{ij}} d\sigma_{ij} > 0$$

whereas *unloading* is given if

$$f < 0$$

$$\text{or} \quad f = 0 \quad \text{and} \quad \frac{\partial f}{\partial \sigma_{ij}} d\sigma_{ij} < 0 \quad .$$

The case $\frac{\partial f}{\partial \sigma_{ij}} d\sigma_{ij} = 0$ constitutes the so-called neutral loading. At loading ε_{ij}^p is varied, i.e. $d\varepsilon_{ij}^p \neq 0$, and the condition

$$\mathrm{d}f = \frac{\partial f}{\partial \sigma_{ij}} \mathrm{d}\sigma_{ij} + \frac{\partial f}{\partial \varepsilon_{ij}^p} \mathrm{d}\varepsilon_{ij}^p = 0 \tag{4.1}$$

(consistency condition) guarantees that the yield surface is carried behind the moving stress point. The direction of plastic strain increment $d\varepsilon_{ij}^p$ is given by an additional function $g(\sigma_{ij})$, the plastic potential, in the following way:

$$d\varepsilon_{ij}^p = \lambda \frac{\partial g}{\partial \sigma_{ij}} \qquad (4.2)$$

Equ. (4.2) is called the flow rule. λ is obtained by introducing (4.2) into (4.1) as

$$\lambda = -\frac{\dfrac{\partial f}{\partial \sigma_{kl}}}{\dfrac{\partial f}{\partial \varepsilon_{pq}^p} \dfrac{\partial g}{\partial \sigma_{pq}}} d\sigma_{kl} \quad .$$

Finally we have for loading

$$
\begin{aligned}
d\varepsilon_{ij} &= d\varepsilon_{ij}^e + d\varepsilon_{ij}^p \\
&= \left[E_{ijkl} - \frac{\dfrac{\partial f}{\partial \sigma_{kl}} \dfrac{\partial g}{\partial \sigma_{ij}}}{\dfrac{\partial f}{\partial \varepsilon_{pq}^p} \dfrac{\partial g}{\partial \sigma_{pq}}} \right] d\sigma_{kl}
\end{aligned}
$$

and for unloading

$$d\varepsilon_{ij} = E_{ijkl} d\sigma_{kl} \quad .$$

The special case $f = g$ is called 'normality condition' or 'associated flow rule'.

4.6 Hypoplasticity

Elastoplastic and hypoplastic equations are both of the general form

$$\overset{\circ}{\mathbf{T}} = \mathbf{h}(\mathbf{T}, \mathbf{D}) \quad .$$

Starting from the fact that every function $\mathbf{h}(\mathbf{T}, \mathbf{D})$ can be represented according to the general representation theorem,

$$
\begin{aligned}
\mathbf{h}(\mathbf{T}, \mathbf{D}) =\ & \psi_1 \mathbf{1} + \psi_2 \mathbf{T} + \psi_3 \mathbf{D} + \psi_4 \mathbf{T}^2 + \psi_5 \mathbf{D}^2 + \psi_6 (\mathbf{TD} + \mathbf{DT}) \\
& + \psi_7 (\mathbf{TD}^2 + \mathbf{D}^2\mathbf{T}) + \psi_8 (\mathbf{T}^2\mathbf{D} + \mathbf{DT}^2) + \psi_9 (\mathbf{T}^2\mathbf{D}^2 + \mathbf{D}^2\mathbf{T}^2)
\end{aligned}
$$

(ψ_i are scalar functions of invariants and joint invariants of $\mathbf{T}$ and $\mathbf{D}$), the experiment was undertaken to find such a unique function which appropriately describes the mechanical properties of soils [37]. In order to avoid the shortcomings of hypoelasticity, this function has to be non-linear in $\mathbf{D}$. On the other hand, it should be homogeneous of the first degree in $\mathbf{D}$ in order to describe rate-independent materials, and homogeneous in $\mathbf{T}$ in order to describe proportional stress-paths in case of proportional strain paths. Therefore, the design of such a function had to proceed along the above stated representation theorem and some general mathematical restrictions:

-    non-linearity in $\mathbf{D}$
-    homogeneity in $\mathbf{D}$ and $\mathbf{T}$

with avoidance of any recourse to notions from the theory of elastoplasticity such as yield functions, decomposition of strain etc.

This experiment (every theory is, virtually, an experiment) was more or less successful, as by trial and error a function was found which was able to describe many aspects of soil behaviour. Thus, a new approach to constitutive modelling was created. The name 'hypoplastic' equation fits very well, as the relation between hypoplasticity and (elasto)plasticity is the same as the one between hypoelasticity and elasticity: The theories with "hypo" do not use a potential. It should be mentioned that DAFALIAS [7] coined the term hypoplasticity earlier for something else, which can be considered as a general case of what we call hypoplasticity.

Let us now have a look at some hypoplastic equations. Most of them consist of 4 tensorial terms (so-called tensor generators) combined together with 4 material parameters C_1, C_2, C_3 and C_4, e.g. [79]:

$$\overset{\circ}{\mathbf{T}} = C_1(\operatorname{tr}\mathbf{T})\mathbf{D} + C_2\frac{\operatorname{tr}(\mathbf{TD})}{\operatorname{tr}\mathbf{T}}\mathbf{T} + C_3\frac{\mathbf{T}^2}{\operatorname{tr}\mathbf{T}}\sqrt{\operatorname{tr}\mathbf{D}^2} + C_4\frac{\mathbf{T}^{*2}}{\operatorname{tr}\mathbf{T}}\sqrt{\operatorname{tr}\mathbf{D}^2} \qquad (4.3)$$

with the deviatoric stress $\mathbf{T}^*$ defined as

$$\mathbf{T}^* = \mathbf{T} - \frac{1}{3}(\operatorname{tr}\mathbf{T})\mathbf{1} \quad .$$

An alternative representation of hypoplastic constitutive equations is to summarize the linear terms by $\mathbf{LD}$, with $\mathbf{L}$ being a linear operator applied to $\mathbf{D}$, and the non-linear terms by $\mathbf{N}|\mathbf{D}|$ with $|\mathbf{D}| := \sqrt{\operatorname{tr}\mathbf{D}^2}$. Then, a hypoplastic equation assumes the general form

$$\overset{\circ}{\mathbf{T}} = \mathbf{LD} + \mathbf{N}|\mathbf{D}| \qquad (4.4)$$

or

$$\overset{\circ}{T}_{ij} = L_{ijkl}D_{kl} + N_{ij}|\mathbf{D}| \quad .$$

The components L_{ijkl} and N_{ij} depend on the actual stress and can easily be numerically determined for a given constitutive relation $\overset{\circ}{\mathbf{T}} = \mathbf{h}(\mathbf{T},\mathbf{D})$: For any given values of the indices, say $k = k^*$ and $l = l^*$, we set $D_{k^*l^*} = 1$, else $D_{kl} = 0$, and obtain from the constitutive relation $\overset{\circ}{T}_{ij}^+$. Then we set $D_{k^*l^*} = -1$ and obtain $\overset{\circ}{T}_{ij}^-$. Subsequently we obtain

$$L_{ijk^*l^*} = \frac{1}{2}\left(\overset{\circ}{T}_{ij}^+ - \overset{\circ}{T}_{ij}^-\right)$$

$$N_{ij} = \frac{1}{2}\left(\overset{\circ}{T}_{ij}^+ + \overset{\circ}{T}_{ij}^-\right) \quad .$$

Note that N_{ij} is independent of k^* and l^*, as it does not depend on $\mathbf{D}$. L_{ijkl} is not necessarily symmetric in the sense $L_{ijkl} \neq L_{klij}$. For the case of equ. 4.3 however, L_{ijkl} is symmetric in the aforementioned sense.

Another expression for constitutive relations of the type (4.4) is

$$\overset{\circ}{\mathbf{T}} = \mathbf{H}\mathbf{D}$$

with

$$\mathbf{H} := \mathbf{L} + \mathbf{N} \otimes \mathbf{D}^0 \quad .$$

Herein, $\mathbf{D}^0$ is the normalized stretching, i.e. $\mathbf{D}^0 := \dfrac{\mathbf{D}}{|\mathbf{D}|}$. In index notation the stiffness matrix $\mathbf{H}$ can be expressed as:

$$H_{ijkl} = L_{ijkl} + N_{ij}D^0_{kl} \quad .$$

Several equations with only 4 material parameters C_1, C_2, C_3, C_4 could be found [34, 76, 78, 79, 39] which were capable to describe

- the triaxial test as characterized by a stiffness decreasing down to zero at the limit state and a corresponding volumetric strain curve exhibiting first contractancy and then dilatancy

- incrementally non-linear behaviour, i.e. unloading stiffness much larger than at loading

- realistic asymptotic properties (referring to proportional paths).

However, the void ratio was not taken into account, and, therefore, such simple hypoplastic constitutive models were not capable of describing the difference of friction angle and stiffness between dense and loose samples, or the decrease of the peak friction angle to the residual one with increasing strain (softening). But this was also not expected from such simple constitutive models. To achieve this, more recent versions have been elaborated in Karlsruhe [3, 16, 41, 77, 84], where several tensorial terms are multiplied with scalar factors which aim to model the influence of density and stress level as well as the transition to the so-called critical state. Of course, such factors increase the intricacy of the models.

Hypoplastic constitutive relations are directly presented without reference to any sort of surfaces in stress space. However, various surfaces can be derived from a hypoplastic equation, as will be explained in Chapter 5.

4.7 Response envelopes

The response envelopes introduced by GUDEHUS [15] to model axisymmetric response are very useful in discussing properties of constitutive models.

With reference to a given axisymmetric stress state we can apply all possible axisymmetric stretchings $\mathbf{D}$ with $|\mathbf{D}| = 1$ by setting

$$\mathbf{D} = \begin{pmatrix} -\sin \alpha & 0 & 0 \\ 0 & -\cos \alpha / \sqrt{2} & 0 \\ 0 & 0 & -\cos \alpha / \sqrt{2} \end{pmatrix}$$

with $0° \leq \alpha \leq 360°$. The corresponding polar representation of the obtained responses $\overset{\circ}{\mathbf{T}}$ is called the response envelope.

GUDEHUS showed that for linear relations between $\mathbf{T}$ and $\mathbf{D}$ (e.g. for hypoelasticity) response envelopes are ellipses: For the aforementioned $\mathbf{D}$-tensor a linear constitutive equation $\overset{\circ}{\mathbf{T}} = \mathbf{f}(\mathbf{D})$ assumes the form

$$\dot{\sigma}_1 = a_{11} \sin \alpha + a_{12} \cos \alpha$$
$$\dot{\sigma}_2 = a_{21} \sin \alpha + a_{22} \cos \alpha$$

which is the parametric representation of an ellipse in the $\dot{\sigma}_1$-$\dot{\sigma}_2$-space. For hypoplasticity the response envelopes are also ellipses, but the reference stress state is no more the centre of the ellipse. Thus, limit states can be modelled:

It is interesting to see in terms of response envelopes that there are several ways to model limit states (see Fig. 4.4, 4.6). Each of them should provide at least one van-

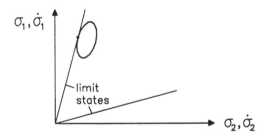

Figure 4.4: Response envelope at limit state (hypoplasticity)

ishing stress rate. With respect to response envelopes this means that the reference stress state must coincide with at least one point of the periphery of the response envelope[1]. For hypoplastic and elastoplastic relations this is the case for only one

---

[1]The condition that the response envelope be tangential to the limit surface reads $\text{tr}\left[(\mathbf{L}^{-1}[\mathbf{T}])(\mathbf{L}^{-1}[\mathbf{N}])\right] = 0$ (J. WALLNER, private communication).

stretching **D**. An alternative way to model limit states (i.e. states with vanishing stress rate) could be obtained with response envelopes shrinked to a point. However, this approach would have the disadvantage that the limit state cannot be abandoned through unloading — unless another response is defined for unloading. This would lead to a sort of discontinuous response which is typical for some 'hypoelastic' constitutive equations, as they have been suggested by DAVIS & MULLENGER [9] and others.

It is worth noting that in elastoplasticity the response envelopes are composed of several ellipses (corresponding each to a linear relation holding for loading or unloading) such that the envelope, as a whole, is continuous but not smooth (see Fig. 4.5). At the limit state one of the ellipses degenerates to a straight line (see Fig. 4.6).

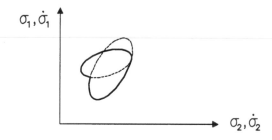

Figure 4.5: Response envelope of an elastoplastic relation

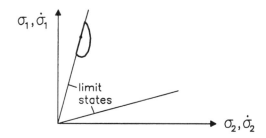

Figure 4.6: Response envelope at limit state (elastoplasticity)

4.8 Numerical simulation of element tests

Whereas constitutive equations relate stresses with strains, in the laboratory we only can measure forces and displacements. Thus, in order to check a constitutive relation, we need tests with homogeneous (i.e. constant) distribution of stress and strain within the sample. If homogeneity is given, then we can easily obtain stresses and strains from boundary forces and displacements. For inhomogeneously deformed samples the distribution of stresses and strains within the sample can only be inferred with some highly sophisticated methods using x-rays or other sorts of transparent samples), and we are only able to numerically calculate the stress and strain fields by means of the constitutive equation we want to check. Thus, constitutive equations have to be based on element tests, i.e. tests with homogeneous sample deformation. In the Russian literature element tests are called 'zero-dimensional' tests. Such tests are connected with extraordinarily high experimental difficulties.

Despite all efforts, experimentalists have to admit that in the course of a test an inhomogeneous deformation will — soon or late — set on, no matter what measures have been undertaken to prevent this. In other words, inhomogeneous deformation is inevitable. This fact adds to soil mechanics similar difficulties as the transition from laminar to turbulent flow in hydromechanics does. Thus, we see that with respect to laboratory tests aiming to support constitutive modelling, (i) we have to spend provisions to achieve homogeneous deformation, and (ii) even so the homogeneous deformation can only be realized for part of the test duration. In other words, nature allows us to look upon homogeneously deformed samples only within a time-window. At that, we do not know exactly the boundaries of this window. Unfortunately, it is a common practice in soil mechanics to ignore inhomogeneous deformation and to evaluate tests as if they were homogeneous. The results are, of course, questionable.

How can we obtain simulations of laboratory element tests by using an equation of the rate type? First, we have to start from a known stress state. If the test to be simulated has kinematical boundary conditions, then the stretching $\mathbf{D}$ is known, e.g. in case of the oedometer test all but one components of $\mathbf{D}$ are equal to zero and the only non-vanishing component corresponds to the rate of compression (cf. exercise 4 in chapter 6). With knowledge of $\mathbf{T}$ and $\mathbf{D}$ the constitutive equation $\overset{\circ}{\mathbf{T}} = \mathbf{h}(\mathbf{T}, \mathbf{D})$ makes possible to evaluate $\overset{\circ}{\mathbf{T}}$. Multiplying $\overset{\circ}{\mathbf{T}}$ with a sufficiently small time step Δt gives $\Delta \mathbf{T} \approx \overset{\circ}{\mathbf{T}} \Delta t$. The new stress state is then obtained to $\mathbf{T} + \Delta \mathbf{T}$. This process can be continued and corresponds to a numerical integration of the evolution equation $\overset{\circ}{\mathbf{T}} = \mathbf{h}(\mathbf{T}, \mathbf{D})$ (so-called EULER-forward integration). The procedure is a little more difficult if not all of the boundary conditions are of the kinematic type. In case of a static boundary condition (e.g. $\sigma_2 = \sigma_3 = $ const for triaxial test), the component D_2 of $\mathbf{D}$ must be determined by solving the algebraic equation $\dot{\sigma}_2(D_2) = 0$ (cf. exercise 5 in chapter 6).

A program to simulate element tests can be downloaded from
`ftp://ftp.uibk.ac.at/pub/uni-innsbruck/igt/sources/`

4.9   Calibration

A constitutive relation is of no use if the involved material parameters cannot be adapted to a particular material. The values of these parameters constitute the identity card of this material with respect to a particular constitutive model. Moreover, a particular parameter is useless unless it is embedded within a constitutive model. E.g. the notion 'viscosity' is unclear unless it is embedded within a NEWTON-type constitutive equation, say $\tau = \mu\dot{\varepsilon}$. The process of the determination of the values of the parameters of a constitutive model is called "calibration" or "parameter identification". In many publications on constitutive models the calibration is simply omitted as being not worth mentioning. In fact it is a task which can take up to several months of work! Considering hypoplastic constitutive equations, the calibration is straightforward by fitting the equation to the outcomes of one or several (say triaxial) tests [36]. If we know the values of the strain and stress increments at a particular stress state from experiments, the only remaining unknowns in the constitutive equation are the material constants. Thus, we have to solve a system of four linear equations.

A possible procedure for calibration of the constitutive equ. (4.3) on the basis of a triaxial test (see Fig. 4.7) is as follows:

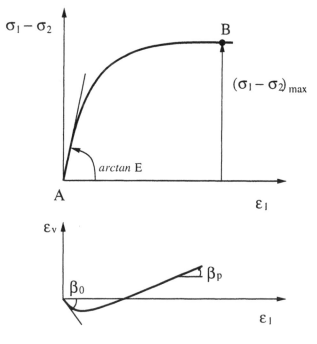

Figure 4.7: Results of a triaxial test

With

$$-\mathbf{T} = \begin{pmatrix} \sigma_1 & 0 & 0 \\ 0 & \sigma_2 & 0 \\ 0 & 0 & \sigma_2 \end{pmatrix} \quad , \quad \mathbf{D} = \begin{pmatrix} \dot{\epsilon}_1 & 0 & 0 \\ 0 & \dot{\epsilon}_2 & 0 \\ 0 & 0 & \dot{\epsilon}_2 \end{pmatrix}$$

we take from Fig. 4.7 the values

$$E_A, \beta_A, \beta_B \quad \text{and} \quad (\sigma_1 - \sigma_2)_{max}$$

and interprete them as follows in terms of $\mathbf{T}$ and $\mathbf{D}$:

$$\beta_A = \arctan\left(\frac{\dot{\epsilon}_v}{\dot{\epsilon}_1}\right)_A = \arctan\left(\frac{\dot{\epsilon}_1 + 2\dot{\epsilon}_2}{\dot{\epsilon}_1}\right)_A = \arctan\left(1 + 2\frac{\dot{\epsilon}_2}{\dot{\epsilon}_1}\right)_A \qquad (4.5)$$

$$\beta_B = \arctan\left(\frac{\dot{\epsilon}_v}{\dot{\epsilon}_1}\right)_B = \arctan\left(\frac{\dot{\epsilon}_1 + 2\dot{\epsilon}_2}{\dot{\epsilon}_1}\right)_B = \arctan\left(1 + 2\frac{\dot{\epsilon}_2}{\dot{\epsilon}_1}\right)_B$$

The indices A and B denote that the corresponding quantities are obtained at the points A and B of Fig. 4.7.

Since the considered experiment is a conventional triaxial compression, the component $\dot{\epsilon}_1$ must be negative. The value (modulus) of $\dot{\epsilon}_1$, i.e. $|\dot{\epsilon}_1|$, is not important, as the constitutive equation is rate-independent. Thus, we can set $\dot{\epsilon}_1 = -1$. It then follows

$$\dot{\epsilon}_2 = \frac{1}{2}(1 - \tan\beta) \quad ,$$

and we can evaluate the tensor $\mathbf{D}$ at the points A and B:

$$\mathbf{D}_{A/B} = \begin{pmatrix} -1 & 0 & 0 \\ 0 & \frac{1}{2}(1 - \tan\beta_{A/B}) & 0 \\ 0 & 0 & \frac{1}{2}(1 - \tan\beta_{A/B}) \end{pmatrix}$$

We also know the stress $\mathbf{T}$ at points A and B:

With $\sigma_{1B} = \sigma_2 + (\sigma_1 - \sigma_2)_{max}$ we have

$$-\mathbf{T}_A = \begin{pmatrix} \sigma_2 & 0 & 0 \\ 0 & \sigma_2 & 0 \\ 0 & 0 & \sigma_2 \end{pmatrix}$$

$$-\mathbf{T}_B = \begin{pmatrix} \sigma_{1B} & 0 & 0 \\ 0 & \sigma_2 & 0 \\ 0 & 0 & \sigma_2 \end{pmatrix} \quad .$$

Finally we know the stress rate $\overset{\circ}{\mathbf{T}} = \dot{\mathbf{T}}$ at the points A and B. With $\dot{\sigma}_1 = E\dot{\epsilon}_1$ and $\dot{\epsilon}_1 = -1$ we have

$$\dot{\mathbf{T}}_A = \begin{pmatrix} -E & 0 & 0 \\ 0 & 0 & 0 \\ 0 & 0 & 0 \end{pmatrix} \quad , \quad \dot{\mathbf{T}}_B = \begin{pmatrix} 0 & 0 & 0 \\ 0 & 0 & 0 \\ 0 & 0 & 0 \end{pmatrix} \quad .$$

To determine the material constants C_1, C_2, C_3, C_4 we write the following scalar components of the constitutive equation (4.3):

$$\begin{aligned} \dot{\sigma}_1(A) &= -E \\ \dot{\sigma}_2(A) &= 0 \\ \dot{\sigma}_1(B) &= 0 \\ \dot{\sigma}_2(B) &= 0 \end{aligned}$$

and obtain a system of 4 algebraic equations for the unknowns C_1, C_2, C_3, C_4. This system can be numerically solved. Since σ_{1B} is related to the friction angle φ by

$$\frac{\sigma_{1B}}{\sigma_2} = \frac{1 + \sin\varphi}{1 - \sin\varphi}$$

we can say that the above stated calibration procedure is a method how to obtain the material parameters C_1, C_2, C_3, C_4 from the measured values β_A, β_B, E and φ. More recently other methods of calibration have been proposed for more advanced versions of the hypoplastic constitutive equation (see section 4.12).

4.10   Dilatancy and pore pressure

The tendency of soils to contract or dilate at shearing is known as a peculiar feature of granular materials. Some people believe that dilatancy can be modelled even with a linear-elastic material provided POISSON's ratio is properly chosen. However, this is not true, as in linear elastic materials the hydrostatic and the deviatoric stresses and strains are completely uncoupled. It is in this coupling where dilatancy resides, since it means that the volumetric strain is affected by deviatoric stress and vice versa. In the course of conventional triaxial tests with dense sand, an initial contractancy (or negative dilatancy) is followed by dilatancy. The initial contractancy is often attributed to compressibility since the volumetric decrease is accompanied with an increase of the hydrostatic stress. However, the initial volume decrease occurs also in tests with constant hydrostatic stress (tr $\mathbf{T}$ = const), so that "contractancy" appears to be a suitable name.

A reference is often made to *the* angle of dilatancy. However, this angle is by no means a material constant. Much more it depends on the deformation mode and stage, on the density and on the stress level. We should get rid of the preoccupation that the dilatancy angle as well as the friction angle are material constants.

Let us now consider suppressed dilatancy (or contractancy) in case of undrained deformation of water saturated soil. An often cited expression is "the soil wants to decrease its volume but it cannot; consequently the hydrostatic effective stress is decreased". Of course, such an explanation is not very satisfactory. A much better approach is to consider pore water as imposing the internal constraint of incompressibility. Then, the porewater pressure u is constitutively indeterminate and can only be determined from static boundary conditions. The constitutive equation is now changed to

$$\overset{\circ}{\mathbf{T}} = \mathbf{h}(\mathbf{T}, \mathbf{D}) + \dot{u}\mathbf{1} \quad .$$

Here, $\mathbf{T}$ should be understood as the effective stress. Thus, the pore pressure built up in the course of undrained tests can be easily and realistically modelled by means of hypoplasticity (cf. exercise 7 in chapter 6).

4.11 Cyclic loading, ratcheting, shake-down

Cyclic loading is recognized as one of the most difficult fields in soil mechanics. Elastoplasticity and hypoplasticity bear some inherent deficiencies which become more important in the case of cyclic loading. In the realm of classical elastoplasticity all unloading-reloading cycles are completely elastic, a feature which is not realistic. On the other hand, in (the initial versions of) hypoplasticity the first and subsequent unloading-reloading cycles are identical to the virgin loading-unloading [49]. This shortcoming is called ratcheting effect and is due to the fact that in hypoplasticity the stress is the only memory parameter.

In reality either a gradual transition from plastic to elastic behaviour (so-called shake-down) takes place or deformation increases unbounded with the number of cycles (so-called incremental collapse). Regarding shake-down, experiments show that the behaviour of soil never becomes completely elastic, as every cycle is connected with dissipation of energy, a fact which is modelled in soil dynamics by a fictitious viscous damping.

It turns out that the quality of the modelling of cyclic behaviour depends on whether the stress amplitudes are small or large. If the unloading is continued to the extension side (i.e. the stress deviator changes sign), then the hypoplastic models work satisfactorily. Furthermore, the proper incorporation of barotropy and pyknotropy by the advanced hypoplastic models enables that cyclic shearing produces gradually a high density (i.e. small void ratio) which cannot be exceeded by additional cycles.

A more general representation of the cyclic behaviour in hypoplasticity requires an additional state variable such as a structure tensor [35] which involves the recent deformation history. A 'memory function' [4] or an 'intergranular strain' [51] have been proposed for this purpose.

4.12   Development of hypoplastic equations — a review

Following some early versions [29, 17, 30, 31, 32] the first hypoplastic equation which satisfied the requirements given in Section 4.6 was published in 1985 [33]:

$$\mathring{\mathbf{T}} = C_1 \frac{1}{2}(\mathbf{TD} + \mathbf{DT}) + C_2 \mathrm{tr}\,(\mathbf{TD})\mathbf{1} + \left[C_3 \mathbf{T} + C_4 \frac{\mathbf{T}^2}{\mathrm{tr}\,\mathbf{T}} \right] \sqrt{\mathrm{tr}\,\mathbf{D}^2} \quad.$$

In order to remove some deficiencies, Wu Wei [76] modified the tensorial terms and obtained the aforementioned relation

$$\mathring{\mathbf{T}} = C_1(\mathrm{tr}\,\mathbf{T})\mathbf{D} + C_2 \frac{\mathrm{tr}\,(\mathbf{TD})}{\mathrm{tr}\,\mathbf{T}}\mathbf{T} + C_3 \frac{\mathbf{T}^2}{\mathrm{tr}\,\mathbf{T}}\sqrt{\mathrm{tr}\,\mathbf{D}^2} + C_4 \frac{\mathbf{T}^{*2}}{\mathrm{tr}\,\mathbf{T}}\sqrt{\mathrm{tr}\,\mathbf{D}^2} \quad.$$

Next, he multiplied the non-linear terms (i.e. the last two terms) by the factor I_e which depends on the void ratio e and becomes equal to 1 when $e = e_{\mathrm{crit}}$:

$$I_e = (1 - a)\frac{e - e_{\min}}{e_{\mathrm{crit}} - e_{\min}} + a$$

The material constants C_1, C_2, C_3, C_4 can be determined from the critical state which can be reached in the course of monotonous shearing. He thus succeeded to model pyknotropy. Moreover, by taking into account that the critical void ratio e_{crit} depends on the stress level he also managed to model barotropy [77].

KOLYMBAS, HERLE and VON WOLFFERSDORFF [41] proposed 1995 a hypoplastic constitutive equation where the influence of stress level and void ratio is incorporated into a stress-like variable $\mathbf{S}$ called 'structure tensor' or 'back stress':

$$\mathring{\mathbf{T}} = C_1[\mathrm{tr}\,(\mathbf{T} + \mathbf{S})]\mathbf{D} + C_2 \frac{\mathrm{tr}\,[(\mathbf{T} + \mathbf{S})\mathbf{D}]}{\mathrm{tr}\,(\mathbf{T} + \mathbf{S})}(\mathbf{T} + \mathbf{S}) +$$

$$\left[C_3 \frac{\mathbf{T}^2}{\mathrm{tr}\,\mathbf{T}} + C_4 \frac{\mathbf{T}^{*2}}{\mathrm{tr}\,\mathbf{T}} + C_5 \frac{\mathbf{T}^3}{\mathrm{tr}\,\mathbf{T}^2} + C_6 \frac{\mathbf{T}^{*3}}{\mathrm{tr}\,\mathbf{T}^2} \right] \sqrt{\mathrm{tr}\,\mathbf{D}^2}$$

with

$$\mathbf{S} = \mathbf{S}(e, \mathrm{tr}\,\mathbf{T}) := s_0 \left[1 - \frac{1}{\left(\frac{\mathrm{tr}\,\mathbf{T}_r}{p_0}\right)^\nu \ln\frac{1+e_r}{1+e_0}} \left(\frac{\mathrm{tr}\,\mathbf{T}}{p_0}\right)^\nu \ln\frac{1+e}{1+e_0} \right] \left(\frac{\mathrm{tr}\,\mathbf{T}}{p_0}\right)^\alpha$$

The subsequent research focused on two goals:

- to improve the performance (i.e. the predictive capacity) of the equation for deviatoric directions different from the one corresponding to conventional tri-axial tests

- to make the calibration easier. GUDEHUS required that the determination of the material parameters should be as far as possible performed on the basis of granulometric properties of the soil.

Another recent achievement is the choice of the 4 tensorial terms following some general requirements, e.g. that the obtained critical limit state coincides with some prescribed curves given in the literature [2, 3, 71]. In particular BAUER [2, 3] and VON WOLFFERSDORFF [71] established relations between the material constants C_1, C_2, C_3, C_4 by adapting the deviatoric yield curve[2] to some known yield curves which are well-established in literature (e.g. the one by MATSUOKA-NAKAI). With the constants C_1, C_2 and $C_3 = C_4$ depending on e and $\mathbf{T}$ and introducing the abbreviation

$$\hat{\mathbf{T}} := \frac{\mathbf{T}}{\operatorname{tr} \mathbf{T}}$$

a recent hypoplastic equation assumes the form:

$$\overset{\circ}{\mathbf{T}} = f_b f_e \frac{1}{\operatorname{tr} \hat{\mathbf{T}}^2} \left[F^2 \mathbf{D} + a^2 \hat{\mathbf{T}} \operatorname{tr}(\hat{\mathbf{T}} \mathbf{D}) + f_d a F(\hat{\mathbf{T}} + \hat{\mathbf{T}}^*) \sqrt{\operatorname{tr} \mathbf{D}^2} \right] \tag{4.6}$$

with

$$a := \frac{\sqrt{3}(3 - \sin \varphi_c)}{2\sqrt{2} \sin \varphi_c} \quad,$$

$$F := \sqrt{\frac{1}{8} \tan^2 \psi + \frac{2 - \tan^2 \psi}{2 + \sqrt{2} \tan \psi \cos 3\vartheta} - \frac{1}{2\sqrt{2}} \tan \psi} \quad,$$

$$\tan \psi = \sqrt{3 \operatorname{tr} \hat{\mathbf{T}}^{*2}} \quad, \qquad \cos 3\vartheta = -\sqrt{6} \frac{\operatorname{tr} \hat{\mathbf{T}}^{*3}}{\left[\operatorname{tr} \hat{\mathbf{T}}^{*2} \right]^{3/2}} \quad,$$

$$f_d := \left(\frac{e - e_d}{e_c - e_d} \right)^\alpha \quad,$$

---

[2]i.e. the cross section of the limit surface with a deviatoric plane

$$f_e := \left(\frac{e_c}{e}\right)^\beta \quad ,$$

$$f_b := \frac{h_s}{n}\left(\frac{e_{i0}}{e_{c0}}\right)^\beta \frac{1+e_i}{e_i}\left(\frac{3p}{h_s}\right)^{1-n}\left[3+a^2-a\sqrt{3}\left(\frac{e_{i0}-e_{d0}}{e_{c0}-e_{d0}}\right)^\alpha\right]^{-1} \quad .$$

The void ratios e_i and e_d bound the admissible states in the plane e vs. $p = -\mathrm{tr}\,\mathbf{T}/3$ and depend (together with the critical void ratio e_c) on p according to relation

$$\frac{e_i}{e_{i0}} = \frac{e_c}{e_{c0}} = \frac{e_d}{e_{d0}} = \exp\left[-\left(\frac{3p}{h_s}\right)^n\right] \quad .$$

There are 8 constants in the hypoplastic equation (4.6). They can be easily determined from simple index and/or element tests [23]. The constants are state-independent, at least in a particular pressure range, thus enabling the application of the hypoplastic equation in boundary value problems with pressure and density variation. The critical friction angle φ_c can be estimated from the angle of repose; $e_{d0} \approx e_{min}$, $e_{c0} \approx e_{max}$ and $e_{i0} \approx 1.2 e_{max}$. Following GUDEHUS [16] the limits e_{min} and e_{max} are pressure-dependent.

The parameters for sand are summarized in Tab. 4.1.

Table 4.1: Hypoplastic parameters of Lausitz sand.

φ_c [°]	h_s [MPa]	n	e_{d0}	e_{c0}	e_{i0}	α	β
33	1600	0.19	0.44	0.85	1.0	0.25	1.0

4.13 Relation of hypoplasticity to other theories

Hypoplasticity is not the only attempt to introduce an alternative to "classical" elasto-plasticity. There are other theories with similar aims and with pronounced similarities: The arc-length theory of ILYUSHIN, RIVLIN and PIPKIN, the endochronic theory initiated by VALANIS [69] and the several models created in Grenoble by DARVE, CHAMBON and others. Besides this, hypoplasticity has to be compared with all other elastoplastic theories. Three international competitions in Montreal [86], Grenoble [19] and Cleveland [58] were devoted to the aim to find out which theory was superior to the others. A series of experimental data were communicated to the participants in order to let them calibrate their models. Based on this, the participants were invited to predict some other tests, the outcomes of which were at that time kept secret.

Such competitions appear reasonable, however none of them was conclusive. The reason is that there is no objective measure for the quality of the predictions. And, much more, there is no way to rate the calibration effort versus the quality of results. After all, philosophers found long ago that the value of a theory cannot be measured by theoretical means.

4.14 FEM-implementations of hypoplasticity

One of the main fields of application (but by no means the only one) of constitutive models is the implementation into finite element (FEM) codes. This was also the case with a series of dissertation theses [87, 74, 59, 55, 57, 64, 43, 8, 13, 25].

Despite the fact that hypoplasticity is rather new and cannot resort to the instruments already developed for elastoplasticity (e.g. radial return algorithms), the performance of hypoplastic laws in FEM-calculations and the comparison with measured results is satisfactory [25, 72, 24, 73].

As known, the approximative solutions obtained with FEM are presented as linear combinations of a priori defined shape functions. Thus, the solution is given by a finite number of, usually, nodal displacements and the corresponding stresses. The method itself consists in establishing equilibrium not at every point of a continuous body but only at some selected points, the so-called nodes. This approach can be derived from the field equations of equilibrium, $\partial \sigma_{ij}/\partial x_j + \rho g_i = 0$, by multiplying them with u_i, the field of virtual displacements, and setting the integral over the body equal to zero. The resulting equation is known as the weak form of the equilibrium conditions. Applying the theorem of GAUSS and using appropriate virtual displacements leads to the aforementioned equilibrium conditions for the nodes.

If the material behaviour is path-dependent, then we may not apply the entire load or boundary displacement at once. Instead, we have to decompose the loading action (be it a boundary traction or displacement) into sufficiently small steps and seek to fulfil nodal equilibrium at each step. Moreover, the imposed non-linearity requires to solve a system of non-linear algebraic equations, a task which can be achieved by application of, say, NEWTON's method. The decomposition of the loading process into small steps allows in principle, to solve problems with arbitrary large deformations. Of course, on overtly large deformation of the mesh may impose the necessity of remeshing.

The FEM programme ABAQUS allows to implement particular constitutive equations within the subroutine UMAT. This subroutine has to be programmed in such a way that for given stress $\sigma_{ij}(t)$ and strain $\varepsilon_{ij}(t)$ at time t and for given strain and time increments, $\Delta\varepsilon_{ij}$ and Δt, respectively, it returns the stress $\sigma_{ij}(t + \Delta t)$ and the

Jacobian $\partial \dot{\sigma}_{ij} / \partial \dot{\varepsilon}_{kl}$ at time $t + \Delta t$. Depending on whether the given strain increment is small enough or not, it may be necessary to subdivide it into smaller steps (so called sub-stepping).

4.15   Initial stress

Equation (4.3) is of the rate type, i.e. it is an evolution equation which makes possible to calculate the stress changes due to a given increment of deformation. The initial stress has to be known or assumed. Thus, the problem of determining the stress can only be back-stepped but not entirely solved by means of equations of the rate type. This fact is, of course, not very pleasant, but there is no means how to circumvent it. We also can rely upon the fact that the influence of the initial state fades with increasing length of the history. Besides this fact there are some cases (e.g. one-dimensional consolidation corresponding to sedimentation) where we can determine the initial stress by reasoning.

The problem of determination of the initial stress is traditionally hidden by elasto-plasticity, where it is always tacitly assumed that the initial stress results from the theory of elasticity. The latter has to be applied for a deformation starting from a stress free state: The gravity is 'switched on' and the transition to the deformed state is considered to be elastic. This simplification is reflected in almost all existing finite element codes. It is not realistic for soils.

Chapter 5

Uniqueness and limit loads

5.1 Limit states

A very important property of granular materials is their ability to flow (or yield), i.e. to undergo large deformations without stress change, as soon as the stresses and the void ratio obtain their critical values. This sort of flow should be attributed as 'plastic' flow and distinguished from the flow of fluids. The latter has a pronounced viscous (rate-dependent) character.

Plastic flow occurs as soon as the stress state $\mathbf{T}$ and the strain rate $\mathbf{D}$ fulfil the condition $\mathbf{h}(\mathbf{T}, \mathbf{D}) = \mathbf{0}$. In the theory of elastoplasticity the condition in terms of $\mathbf{T}$ is called the yield (limit) surface, and the condition in terms of $\mathbf{D}$ is called the flow rule.

In elastoplasticity the yield function is the starting point and the mathematical relation connecting strain and stress increments at loading is based upon this yield function. In contrast, it can be shown that a yield function is contained in a hypoplastic formulation $\overset{\circ}{\mathbf{T}} = \mathbf{h}(\mathbf{T}, \mathbf{D})$, i.e. the yield function $f(\mathbf{T})$ can be derived from the constitutive relation. To this purpose we rewrite (following a proposition of DESRUES and CHAMBON [11]) the equation (4.4) in the form

$$\overset{\circ}{\mathbf{T}} = \mathbf{L}(\mathbf{T})[\mathbf{D}] + \mathbf{N}(\mathbf{T})|\mathbf{D}| = \mathbf{L}(\mathbf{T})\left[\mathbf{D} + \mathbf{B}|\mathbf{D}|\right] \quad ,$$

with $\mathbf{L}(\mathbf{T})$ being a matrix operator applied to its tensorial argument. It is obvious that $\mathbf{h}(\mathbf{T}, \mathbf{D}) = \mathbf{0}$ occurs for

$$\mathbf{D}^0 := \mathbf{D}/|\mathbf{D}| = -\mathbf{B} \quad .$$

Consequently, the function $f(\mathbf{T})$ reads

$$f(\mathbf{T}) = \mathrm{tr}\mathbf{B}^2 - 1 \quad ,$$

with $\mathbf{B}$ being a function of $\mathbf{T}$. In other words, the limit surface reads:

$$f(\mathbf{T}) = \mathrm{tr}\mathbf{B}^2 - 1 = 0 \quad .$$

For the constitutive equation (4.3) $\mathbf{B}$ reads as follows[1]:

$$\mathbf{B} = \mathbf{L}^{-1}\mathbf{N} = \frac{C_3\mathbf{T}^2}{C_1(\operatorname{tr}\mathbf{T})^2} + \frac{C_4\mathbf{T}^{*2}}{C_1(\operatorname{tr}\mathbf{T})^2} - \left(\frac{C_2C_3}{C_1}\frac{\operatorname{tr}(\mathbf{T}^3)}{(\operatorname{tr}\mathbf{T})^2} + \frac{C_2C_4}{C_1}\frac{\operatorname{tr}(\mathbf{T}\mathbf{T}^{*2})}{(\operatorname{tr}\mathbf{T})^2}\right)$$
$$\times \frac{\mathbf{T}}{C_1(\operatorname{tr}\mathbf{T})^2 + C_2\operatorname{tr}(\mathbf{T}^2)} \quad .$$

Due to the homogeneity of $\mathbf{h}(\mathbf{T},\mathbf{D})$ (and consequently also of $\mathbf{B}$) in $\mathbf{T}$, the surface $f(\mathbf{T}) = 0$ is a cone with apex at the origin $\mathbf{T} = \mathbf{0}$. The cross section of this cone with the deviatoric plane reveals the influence of the intermediate principal stress, i.e. the yield surface differs from the one determined by the MOHR-COULOMB criterion (where the intermediate principal stress does not play any role).

5.2   Invertibility and controllability

In kinematically controlled tests (such as oedometric test or undrained triaxial test) the stretching $\mathbf{D}$ is prescribed and the stress rate $\overset{\circ}{\mathbf{T}}$ can be uniquely determined by means of the hypoplastic constitutive equation. What about the unique determination of $\mathbf{D}$ when $\overset{\circ}{\mathbf{T}}$ is prescribed? To answer this question of unique invertibility we[2] multiply the equation[3] $\overset{\circ}{\mathbf{T}} = \mathbf{L}\mathbf{D} + \mathbf{N}|\mathbf{D}|$ with the inverse operator $\mathbf{L}^{-1}$ and obtain

$$\mathbf{A} := \mathbf{L}^{-1}\overset{\circ}{\mathbf{T}} = \mathbf{D} + \mathbf{L}^{-1}\mathbf{N}|\mathbf{D}|$$

or

$$\mathbf{D} = \mathbf{A} - \mathbf{B}|\mathbf{D}| \tag{5.1}$$

with $\mathbf{B} := \mathbf{L}^{-1}\mathbf{N}$. With the notation $\mathbf{X} \cdot \mathbf{Y} := \operatorname{tr}(\mathbf{X}\mathbf{Y})$ we obtain from (5.1):

$$\mathbf{D} \cdot \mathbf{D} = (\mathbf{A} - \mathbf{B}|\mathbf{D}|) \cdot (\mathbf{A} - \mathbf{B}|\mathbf{D}|) = \mathbf{A} \cdot \mathbf{A} - 2\mathbf{A} \cdot \mathbf{B}|\mathbf{D}| + \mathbf{B} \cdot \mathbf{B}|\mathbf{D}|^2 \quad . \tag{5.2}$$

Noting that $\mathbf{D} \cdot \mathbf{D} \equiv |\mathbf{D}|^2$ we observe that (5.2) is a quadratic equation for $x := |\mathbf{D}|$. Its solution reads

$$x_{1/2} = \frac{2\mathbf{A} \cdot \mathbf{B} \pm \sqrt{4(\mathbf{A} \cdot \mathbf{B})^2 - 4\mathbf{A} \cdot \mathbf{A}(\mathbf{B} \cdot \mathbf{B} - 1)}}{2(\mathbf{B} \cdot \mathbf{B} - 1)} \quad .$$

---

[1] private communication by J. NADER
[2] The author is indebted to Dr. P. WAGNER, Innsbruck, for many valuable suggestions to this section
[3] for simplicity, the brackets in $\mathbf{L}[\mathbf{D}]$ are omitted

Since x is a modulus, only a solution $x > 0$ is meaningful. Moreover, in order to obtain a unique solution we have to require that only one solution is positive, i.e. $x_1 \cdot x_2 < 0$:

$$x_1 \cdot x_2 = \frac{4(\mathbf{A} \cdot \mathbf{B})^2 - 4(\mathbf{A} \cdot \mathbf{B})^2 + 4\mathbf{A} \cdot \mathbf{A}(\mathbf{B} \cdot \mathbf{B} - 1)}{4(\mathbf{B} \cdot \mathbf{B} - 1)^2} = \frac{\mathbf{A} \cdot \mathbf{A}}{\mathbf{B} \cdot \mathbf{B} - 1} < 0$$

Since $\mathbf{A} \cdot \mathbf{A} > 0$ we infer (cf. [49]) that invertibility is given for $\mathbf{B} \cdot \mathbf{B} - 1 < 0$, i.e. for all stress states inside the limit surface $\mathbf{B} \cdot \mathbf{B} - 1 = 0$, as already pointed by CHAMBON [6].

A more subtle question on unique solutions of the constitutive equation arises if some (say k) components of $\mathbf{D}$ and $6 - k$ components of $\mathring{\mathbf{T}}$ are prescribed, and the remaining components have to be determined[4]. The existence of a unique solution of this problem (which corresponds e.g. to the conventional triaxial test with the mixed conditions $D_{11} = -1$ in axial direction and $\mathring{T}_{22} = \mathring{T}_{33} = 0$ in lateral directions) is called controllability [52]. For simplicity we consider $\mathbf{D}$ and $\mathring{\mathbf{T}}$ as column or row vectors $\mathbf{x}$ and $\mathbf{y}$, i.e. we take $x_1 := D_{11}$, $x_2 := D_{12}$, ... and similarly $y_1 := \mathring{T}_{11}$, $y_2 := \mathring{T}_{12}$, ...

The selection of the independent and dependent components can be accomplished by the partition matrices $\mathbf{P}$ and $\mathbf{Q}$, the components of which vanish for $i \neq j$. Their diagonal components are either 1 or 0. $\mathbf{P}$ and $\mathbf{Q}$ are related by $\mathbf{P} + \mathbf{Q} = \mathbf{1}$, with $\mathbf{1}$ being the unit matrix. E.g.

$$\mathbf{P} = \begin{pmatrix} 1 & 0 & 0 & 0 & 0 & 0 \\ 0 & 0 & 0 & 0 & 0 & 0 \\ 0 & 0 & 0 & 0 & 0 & 0 \\ 0 & 0 & 0 & 1 & 0 & 0 \\ 0 & 0 & 0 & 0 & 0 & 0 \\ 0 & 0 & 0 & 0 & 0 & 1 \end{pmatrix}, \quad \mathbf{Q} = \begin{pmatrix} 0 & 0 & 0 & 0 & 0 & 0 \\ 0 & 1 & 0 & 0 & 0 & 0 \\ 0 & 0 & 1 & 0 & 0 & 0 \\ 0 & 0 & 0 & 0 & 0 & 0 \\ 0 & 0 & 0 & 0 & 1 & 0 \\ 0 & 0 & 0 & 0 & 0 & 0 \end{pmatrix}.$$

We can now obtain the independent (or controlling) variable $\mathbf{X}$ of a problem with mixed conditions as

$$\mathbf{X} = \mathbf{Q}\mathbf{x} + \mathbf{P}\mathbf{y} \tag{5.3}$$

and, likewise, the dependent variable $\mathbf{Y}$ as:

$$\mathbf{Y} = \mathbf{P}\mathbf{x} + \mathbf{Q}\mathbf{y} \tag{5.4}$$

---

[4]As NOVA points out, the most general case of test control (e.g. a test with $T_1 + T_2 + T_3 = \text{const}$, $D_2 = D_3, D_1 = 1$) is obtained if we replace $\mathring{\mathbf{T}}$ by $\mathring{\mathbf{T}}' := \mathbf{S}\mathring{\mathbf{T}}$ and $\mathbf{D}$ by $\mathbf{D}' := \mathbf{S}^{-1}\mathbf{D}$ with some appropriately chosen non-singular matrix $\mathbf{S}$. Obviously, $\mathring{\mathbf{T}}'$ and $\mathbf{D}'$ are energy-conjugated in the sense that $\text{tr}(\mathbf{T}\mathbf{D}) = \text{tr}(\mathbf{T}'\mathbf{D}')$ or $\text{tr}\,\mathring{\mathbf{T}}\mathbf{D}) = \text{tr}\,(\mathring{\mathbf{T}}'\mathbf{D}')$.

E.g.

$$\begin{aligned}
\mathbf{X} &= (X_1, X_2, \ldots)^T &= (D_{11}, \mathring{T}_{12}, \mathring{T}_{13}, D_{22}, \ldots)^T \\
\mathbf{Y} &= (Y_1, Y_2, \ldots)^T &= (\mathring{T}_{11}, D_{12}, D_{13}, \mathring{T}_{22}, \ldots)^T
\end{aligned}$$

Inverting the system of equations (5.3) and (5.4) and using $(1 - 2\mathbf{P})^{-1} = (1 - 2\mathbf{P})$ we obtain

$$\mathbf{x} = \mathbf{QX} + \mathbf{PY} \qquad (5.5)$$

$$\mathbf{y} = \mathbf{PX} + \mathbf{QY} \qquad (5.6)$$

Inserting (5.5) into the constitutive equation $\mathbf{y} = \mathbf{h}(\mathbf{x})$ or $\mathbf{y} - \mathbf{h}(\mathbf{x}) = 0$ we obtain an implicit relation between $\mathbf{X}$ and $\mathbf{Y}$:

$$\mathbf{F}(\mathbf{X}, \mathbf{Y}) := \mathbf{PX} + \mathbf{QY} - \mathbf{h}(\mathbf{QX} + \mathbf{PY}) = 0 \quad . \qquad (5.7)$$

A unique determination of $\mathbf{Y}$ from (5.7) (i.e. controllability) is possible if $\det(\partial \mathbf{F}/\partial \mathbf{Y}) = \det(\partial F_i/\partial Y_j) \neq 0$. This means that

$$\det \left(\mathbf{Q} - \mathbf{P}\frac{d\mathbf{h}}{d\mathbf{x}} \right) \neq 0 \quad .$$

$\dfrac{\partial \mathbf{h}}{\partial \mathbf{x}}$ is the stiffness matrix. For a hypoplastic constitutive equation $\mathbf{h}(\mathbf{x}) := \mathbf{Lx} + \mathbf{N}|\mathbf{x}|$ the stiffness matrix reads

$$\mathbf{H}(\mathbf{x}) := \frac{d\mathbf{h}}{d\mathbf{x}} = \mathbf{L} + \mathbf{N} \otimes \frac{\mathbf{x}}{|\mathbf{x}|} = \mathbf{L} + \mathbf{N} \otimes \mathbf{x}^0 \quad , \qquad (5.8)$$

i.e. the stiffness matrix depends on the direction of $\mathbf{x}$. This is to be contrasted with elastoplastic formulations where

$$\mathbf{H} = \begin{cases} \mathbf{L}_{elastic} & \text{for unloading or inside the yield surface} \\ \mathbf{L}_{plastic} & \text{for loading} \end{cases}$$

The application of the operator $\mathbf{P}$ to $\mathbf{H}(\mathbf{x})$ selects from $\mathbf{H}(\mathbf{x})$ only those rows which have a non-vanishing $\mathbf{P}$-component. E.g. for[5]

$$\mathbf{P} = \begin{pmatrix} 1 & 0 & 0 \\ 0 & 1 & 0 \\ 0 & 0 & 0 \end{pmatrix} \quad , \quad \mathbf{Q} = 1 - \mathbf{P} = \begin{pmatrix} 0 & 0 & 0 \\ 0 & 0 & 0 \\ 0 & 0 & 1 \end{pmatrix}$$

we obtain

$$\mathbf{PH}(\mathbf{x}) = \begin{pmatrix} H_{11} & H_{12} & H_{13} \\ H_{21} & H_{22} & H_{23} \\ 0 & 0 & 0 \end{pmatrix} \quad .$$

---

[5] for simplicity only 3×3 matrices are considered here

Subtracting this from $\mathbf{Q}$ we obtain

$$\mathbf{Q} - \mathbf{PH}(\mathbf{x}) = - \begin{pmatrix} H_{11} & H_{12} & H_{13} \\ H_{21} & H_{22} & H_{23} \\ 0 & 0 & -1 \end{pmatrix} \quad ,$$

such that

$$\det(\mathbf{Q} - \mathbf{PH}(\mathbf{x})) = \det \begin{pmatrix} H_{11} & H_{12} \\ H_{21} & H_{22} \end{pmatrix} \quad .$$

We thus see that controllability is given if the determinants of all conceivable symmetric minors of $\mathbf{H}(\mathbf{x})$ are positive. This is the case if $\mathbf{H}(\mathbf{x})$ fulfils the condition[6]

$$\mathbf{x} \cdot \mathbf{H}(\mathbf{x})\mathbf{x} > 0 \quad \text{for} \quad \forall \mathbf{x} \neq \mathbf{0} \quad . \tag{5.9}$$

Let $\hat{\mathbf{H}}$ be a symmetric minor of $\mathbf{H}$, and let $\mathbf{H}^s$ be the symmetric part of $\mathbf{H}$. According to the theorem of OSTROWSKY and TAUSSKY [54] cited in [26] condition (5.9) implies that $\det \hat{\mathbf{H}} \geq \det \hat{\mathbf{H}}^s > 0$. Note [7] that $\mathbf{x} \cdot \mathbf{H}(\mathbf{x})\mathbf{x}$ represents the so-called second order work $\text{tr}(\overset{..}{\mathbf{T}}\mathbf{D})$. Thus positive second order work implies controllability. In other words, positive second order work is sufficient (but not necessary) condition for controllability:

$$\text{tr}(\overset{..}{\mathbf{T}}\mathbf{D}) > 0 \qquad \longrightarrow \qquad \text{controllability is guaranteed}$$
$$\text{lack of controllability} \qquad \longrightarrow \qquad \text{tr}(\overset{\circ}{\mathbf{T}}\mathbf{D}) < 0$$

It is interesting to note that, with hypoplastic constitutive equations, the condition $\text{tr}(\overset{..}{\mathbf{T}}\mathbf{D}) = 0$ is in fact encountered *before* the peak. More specifically, the condition $\text{tr}(\overset{\circ}{\mathbf{T}}\mathbf{D}) = 0$ (i.e. vanishing second order work) constitutes a surface in the stress space. Since this surface is connected with possible loss of uniqueness, we call it "bifurcation surface". It can be easily determined if we insert the hypoplastic equation into the equation $\text{tr}(\overset{..}{\mathbf{T}}\mathbf{D}) = 0$. For simplicity we consider only rectilinear extensions, i.e. we restrict the dimensions of column vectors $\mathbf{x}$ and $\mathbf{y}$ to 3. We then obtain:

$$F(\mathbf{x}) = \text{tr}(\overset{..}{\mathbf{T}}\mathbf{D}) = L_{ij}x_i x_j + N_i x_i \cdot |\mathbf{x}| = 0 \tag{5.10}$$

The bifurcation surface is defined as a surface in the stress space. It consists of stress states for which the equation $\text{tr}(\overset{..}{\mathbf{T}}\mathbf{D}) = 0$ possesses only one solution. With $\mathbf{x}^0 := \mathbf{x}/|\mathbf{x}|$ equation (5.10) can be written as

---

[6]The more general case $\mathbf{y} \cdot \mathbf{H}(\mathbf{x})\mathbf{y} > 0$ for $\forall \mathbf{x}, \mathbf{y}$ is not considered here.

[7]A constant matrix $\mathbf{H}$ fulfilling the condition $\mathbf{x} \cdot \mathbf{H}\mathbf{x} \equiv \mathbf{x}^T\mathbf{H}\mathbf{x} > 0$ is called positive definite. All the eigenvalues of $\mathbf{H}^s := \frac{1}{2}(\mathbf{H} + \mathbf{H}^T)$ are positive.

$$\mathbf{x}^0 \cdot (\mathbf{L}\mathbf{x}^0 + \mathbf{N}) = 0 \tag{5.11}$$

This equation is fulfilled in two cases: first for $\mathbf{L}\mathbf{x}^0 + \mathbf{N} = \mathbf{0}$, or $\mathbf{x}^0 = -\mathbf{L}^{-1}\mathbf{N}$, which occurs on the limit state. As an alternative (so-called FREDHOLM's alternative), equ. (5.11) is fulfilled if $\mathbf{x}^0$ is orthogonal to $\mathbf{L}\mathbf{x}^0 + \mathbf{N}$. All stress states for which only one $\mathbf{x}$ fulfils this condition constitute the so-called bifurcation surface. To determine the bifurcation surface we require

$$\nabla F - \lambda \nabla G = 0 \tag{5.12}$$

with

$$G(x) = |x|^2 - 1 = x_1^2 + x_2^2 + x_3^2 - 1 = 0 \tag{5.13}$$

Introducing (5.13) into (5.12) we obtain

$$2L_{(ij)}x_j + N_i + N_k x_k x_i - 2\lambda x_i = 0 \tag{5.14}$$

For $i = 1, 2, 3$ the vectorial equation (5.14) corresponds to three scalar equations. We can search for a point of the bifurcation surface on the deviatoric plane

$$T_1 + T_2 + T_3 = \text{const} \tag{5.15}$$

and on the ray (cf. equation 3.11)

$$\frac{\xi}{\eta} = \text{const} \tag{5.16}$$

The system of 7 algebraic equations (5.10), (5.13), (5.14), (5.15) and (5.16) makes finally possible to determine numerically the 7 unknowns $T_1, T_2, T_3, D_1, D_2, D_3, \lambda$. The analytical representation of the bifurcation surface in terms of $\mathbf{L}$ and $\mathbf{N}$ is too complex.

Corollary 1: As already noted, the theory of elastoplasticity uses (at least) two constant stiffness matrices, one for unloading or inside the yield surface and one (or more, depending on the direction of stress increment) for loading. A known result (see e.g. [48]) in elastoplasticity is that the limit states are defined by $\det(\mathbf{L}_{plastic}) = 0$, whereas a bifurcation (i.e. a non-unique solution of an element test with mixed boundary conditions) may set on if $\det(\mathbf{L}_{plastic}^s) = 0$, where $\mathbf{L}_{plastic}^s$ is the symmetric part[8] of $\mathbf{L}_{plastic}$. We can easily see that these results can also be obtained from the above derivations if we set $\mathbf{N} = \mathbf{0}$ and $\mathbf{L} = \mathbf{L}_{plastic}$. From the definition of limit state, $\overset{\cdot}{\mathbf{T}} = \mathbf{L}\mathbf{D} + \mathbf{N}|\mathbf{D}| = \mathbf{0}$ the limit state condition follows immediately, $\det(\mathbf{L}) = 0$ for $\mathbf{N} = \mathbf{0}$. With $\mathbf{N} = \mathbf{0}$ and with $\mathbf{x}^0 \cdot \mathbf{L}\mathbf{x}^0 = \mathbf{x}^0 \cdot \mathbf{L}^s \mathbf{x}^0$ we obtain from (5.11) the known equation of the bifurcation surface in elastoplasticity, $\det(\mathbf{L}^s) = 0$.

Corollary 2: As can be inferred from Fig. 5.1, negative second order work means negative stiffness. Of course, we should keep in mind that stiffness is a fourth or-

---

[8]Note that the eigenvalues of $\mathbf{L}$ and of $\mathbf{L}^s$ are not identical.

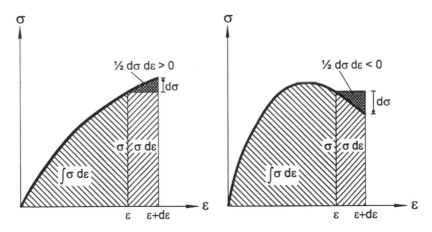

Figure 5.1: Positive (left) and negative (right) second order work

der tensor, $d\sigma_{ij}/d\varepsilon_{kl}$, so that the notion 'negative stiffness' is virtually meaningless. What is here meant is that we obtain negative slope of a graph representing a particular combination of stress components plotted over a particular combination of strain components. If the equation $\mathrm{tr}\,(\overset{\circ}{\mathbf{T}}\mathbf{D}) = 0$ possesses only one solution, this means that there is only one $\mathbf{D} = \mathbf{D}_1$, for which $\mathrm{tr}\,(\overset{\circ}{\mathbf{T}}(\mathbf{D}_1)\mathbf{D}_1) = 0$. Similarly, for $\mathrm{tr}\,(\overset{\circ}{\mathbf{T}}\mathbf{D}) < 0$ several tensors $\mathbf{D}$ can be found that are connected with negative stiffness. If the condition $\mathrm{tr}\,(\overset{\circ}{\mathbf{T}}\mathbf{D}) = 0$ is encountered in the ascending part of the stress-strain curve of a conventional triaxial test, this means that there is a deformation modus (i.e. a specific $\mathbf{D}$) — different than the one corresponding to the homogeneous triaxial deformation — connected with vanishing stiffness.

5.3 Softening

It has been often discussed in soil mechanics, whether softening (see Fig. 5.2) is a material property or not. In this context (and in contrast to the theory of elasto-plasticity) softening is understood as negative stiffness. Traditionally, softening was considered as a principal part of soil behaviour. Later on it became fashionable to deny softening, as being only an apparent effect due to the inhomogeneous deformation of the sample. The view in hypoplasticity is that a large amount of the registered softening is due to the inhomogeneous sample deformation. However, the 'material' softening, i.e. the softening which would be exhibited by a fictitious sample undergoing homogeneous deformation, is also there. We have to admit that the onset of inhomogeneous deformation makes the experimental approach infeasible. We can, however, proceed by reasoning: It is a matter of fact that dense samples have a higher strength (i.e. peak stress deviator) than loose ones. In the course of deformation, dilatancy transforms a sample from dense to loose. Consequently, its strength must decrease and this is material softening.

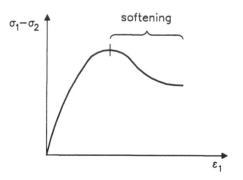

Figure 5.2: Softening in triaxial test deformation

5.4   Shear Bands

A typical pattern of inhomogeneous deformation is the localization of deformation within a narrow zone called shear band (see Fig. 5.3, 5.4).

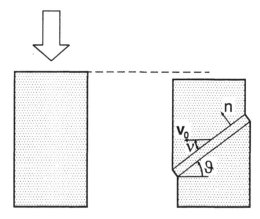

Figure 5.3: Soil sample before and after shear-banding in plane (biaxial) deformation

Such shear bands constitute one of the most fascinating phenomena in geomechanics. Due to the work of DESRUES based on tomography, we know now that also

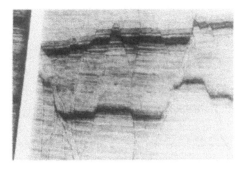

Figure 5.4: Examples of shear bands in nature

apparently non-localized inhomogeneous deformation modes are actually localized. The transition to localized deformation may occur either gradually or suddenly. In the latter case it consists in a drastic change of the deformation direction, as experiments by VARDOULAKIS show. If a constitutive model is capable of realistically describing the material behaviour (i.e. the stiffness) also in this new direction, then it will be possible to predict when and under which inclination a shear band can occur. This ability is not self-evident since many constitutive models (e.g. the elastoplastic ones) are suggested or tested only for some particular fans of deformation directions. It is therefore a good check of constitutive relations to predict the formation of shear bands. This test has been passed by several versions of the hypoplastic relations [30, 35, 40, 60, 64, 65, 82, 83].

The shear band divides the initially homogeneous sample into three parts (see Fig. 5.3), the upper and the lower parts separated by a thin shear zone whose thickness is undetermined. It is realistic to assume that the upper and lower parts do not deform after the spontaneous formation of the shear band, i.e. they behave as rigid bodies (i.e. $\mathbf{D} = \mathbf{0}$, and consequently $\overset{\circ}{\mathbf{T}} = \mathbf{0}$) whereas inside the shear band the motion is described by the velocity gradient $\mathbf{v}_0 \otimes \mathbf{n}$ with $\mathbf{v}_0 := |\mathbf{v}_0|\mathbf{m}$. Since in the rigid parts $\overset{\circ}{\mathbf{T}} = \dot{\mathbf{T}} = \mathbf{0}$ holds true, it follows that the rate of the traction acting upon the discontinuity separating these parts from the shear zone must vanish:

$$\dot{\mathbf{T}}\mathbf{n} = \mathbf{0} \quad . \tag{5.17}$$

Equ. (5.17) is the condition for the spontaneous appearance of a shear band, i.e. a shear band can only appear if this equation possesses a solution. Clearly, $\dot{\mathbf{T}} = \overset{\circ}{\mathbf{T}} + \mathbf{W}\mathbf{T} - \mathbf{T}\mathbf{W}$ depends via the constitutive equation on the motion within the shear zone. This motion is described by $\mathbf{D} = \frac{1}{2}(\mathbf{m} \otimes \mathbf{n} + \mathbf{n} \otimes \mathbf{m})$ and $\mathbf{W} = \frac{1}{2}(\mathbf{m} \otimes \mathbf{n} - \mathbf{n} \otimes \mathbf{m})$ or $D_{ij} = m_{(i}n_{j)} := \frac{1}{2}(m_i n_j + n_i m_j)$, $W_{ij} = m_{[i}n_{j]} := \frac{1}{2}(m_i n_j - n_i m_j)$, where $\mathbf{m}$ and $\mathbf{n}$ are unit vectors.

Introducing the hypoplastic constitutive equation (4.4) into (5.17) yields:

$$\left(L_{ijkl}m_{(k}n_{l)} + N_{ij}\sqrt{(n_km_l)(n_km_l)} + m_{[i}n_{k]}\sigma_{kj} - \sigma_{ik}m_{[k}n_{j]} \right) n_j = 0 \quad (5.18)$$

For plane deformation the unit vectors **m** and **n** can be expressed by means of the angles ϑ and ν. It turns out that the two scalar equations corresponding to (5.18) for $i = 1, 2$ possess a solution for the unknowns ϑ and ν if a 4th degree polynomial in $\sin \nu$ possesses a solution. This can be easily checked by means of STURM chains. Thus, hypoplasticity yields realistic and 'class A' predictions of stress states with the earliest possible onset of shear bands as well as realistic predictions of ϑ and ν [30, 40].

With renamed indices (5.18) transforms to

$$A_{ij}m_j = 0 \tag{5.19}$$

with

$$A_{ij} := n_l(L_{iljk} + L_{ikjl})n_k + N_{ij}|\mathbf{D}| + n_k\sigma_{ki}n_i - n_i\sigma_{jk}n_k - \sigma_{ij} + \sigma_{ik}n_kn_j$$

For the transition to elastoplasticity ($N_{ij} = 0$) the expression in brackets is called the 'acoustic tensor' **A**. Solubility of (5.19) requires that the determinant of A_{ij} vanishes. Thus, $\det(\mathbf{A}) = 0$ is the criterion for shear band formation in elastoplasticity. The condition $\det(\mathbf{A}) = 0$ is called 'loss of strong ellipticity'.

The name 'acoustic tensor' originates from the theory of propagation of elastic waves in anisotropic elastic media (see e.g. [42]). With u_i being the displacement and $\varepsilon_{ij} = u_{(i,j)}$ the strain, we can obtain from momentum balance $\rho\ddot{u}_i = \sigma_{ik,k}$ and constitutive equation $\sigma_{ik} = L_{iklm}\varepsilon_{lm}$ the wave equation

$$\rho\ddot{u}_i = L_{iklm}\frac{\partial^2 u_m}{\partial x_k \partial x_l} \quad .$$

To obtain solutions of the type $u_i = u_{oi}e^{i(\mathbf{kr}-\omega t)}$, the following condition must be fulfilled

$$|L_{iklm}k_kk_l - \rho\omega^2\delta_{im}| = 0 \quad .$$

5.5 Bifurcation modes for 2D and 3D problems

5.5.1 Formulation with finite elements

Considering hypoplastic or elastoplastic materials, initial-boundary-value-problems can be numerically solved with the method of finite elements. Doing so we consider equilibrium of m nodal points and obtain $n = n_d \cdot m$ equations, where n_d is the number of spatial dimensions (i.e. 1 or 2 or 3). Usually we are interested in the time evolution of an equilibrium state as it changes due to e.g. an external loading process. We then consider equations that express the continuation of equilibrium and have the form

$$K_{ij}\dot{x}_j = \dot{y}_i \qquad\qquad i,j = 1, \ldots n \qquad\qquad (5.20)$$

Herein K_{ij} is the incremental (or tangential) stiffness matrix of the considered body, $\dot{x}_j$ are the nodal velocities and $\dot{y}_i$ are the rates of nodal forces due to external volume or surface forces (tractions). Usually the loading process will be controlled by the prescription of the displacement and traction rates for some boundary nodes, whereas the remaining boundary nodes will be free of tractions or displacements. Thus, we re-define n in equ. (5.20) as the number of *unknown* nodal velocities. The vector $\dot{y}_i$ results from quantities controlling the loading process (even if control is purely kinematical, i.e. description of boundary displacement, $\dot{y}_i$ contains non-zero components). The global stiffness matrix K_{ij} is a constant matrix only for linear-elastic materials. For plastic materials, K_{ij} depends on the solution of equ. (5.20), i.e. $K_{ij} = K_{ij}(\dot{x})$. Therefore, equ. (5.20) can only be solved iteratively, say by means of the NEWTON method. The dependency of K_{ij} on $\dot{x}$ for elastoplastic materials is intricate, since the material stiffness is piecemeal linear and the distinction between loading and unloading is based on a series of criteria. In the analysis of bifurcation solutions of (5.20) for elastoplastic materials, the dependence of K_{ij} on $\dot{x}$ is tacitly suppressed and K_{ij} is considered as a constant matrix. In hypoplasticity, the dependence of K_{ij} on $\dot{x}$ is simpler:

$$K_{ij}\dot{x}_j = L_{ij}\dot{x}_j + N_{ip}\sqrt{a_{pkl}\dot{x}_k\dot{x}_l}$$

The constants L_{ij}, N_{ip}, a_{pkl} depend on the hypoplastic constitutive equation and the discretization operations.

5.5.2 Bifurcation modes

In elastoplasticity the non-linearity of (5.20) is simply neglected and K_{ij} is considered as a constant matrix. As K_{ij} is non-symmetric, we have to distinguish between left and right eigenvectors. The right eigenvectors $\mathbf{v}$ are solutions to the problem $\mathbf{K}\mathbf{v} = \lambda\mathbf{v}$, and the left eigenvectors $\mathbf{w}$ are solutions to the problem $\mathbf{K}^T\mathbf{w} = \lambda\mathbf{w}$. Both problems have the same eigenvalues λ_i, but different eigenvectors. Eigenvectors belonging to different eigenvalues are orthogonal, i.e. $\mathbf{v}_i \cdot \mathbf{w}_j = 0$ if $i \neq j$. If we normalize the eigenvectors, we obtain $\mathbf{v}_i \cdot \mathbf{w}_j = \delta_{ij}$. Thus the vectors $\mathbf{v}_i$ (or $\mathbf{w}_j$) can serve as basis to represent any vector $\mathbf{x}$: $\mathbf{x} = \alpha_i\mathbf{v}_i$. Multiplying this equation with $\mathbf{w}_i$ and using the aforementioned orthogonality we obtain $\alpha_i = \mathbf{x} \cdot \mathbf{w}_i$. Thus we have $\mathbf{x} = (\mathbf{x} \cdot \mathbf{w}_i)\mathbf{v}_i$.

Let now $\dot{\mathbf{x}}_0$ be a solution of $\mathbf{K}\dot{\mathbf{x}} = \dot{\mathbf{y}}$. If this is not the unique solution, there must exist a $\dot{\mathbf{x}}_1 \neq \dot{\mathbf{x}}_0$ such that

$$\mathbf{K}\dot{\mathbf{x}}_0 = \dot{\mathbf{y}} \quad ,$$

$$\mathbf{K}\dot{\mathbf{x}}_1 = \dot{\mathbf{y}} \quad ,$$

hence

$$\mathbf{K}(\dot{\mathbf{x}}_1 - \dot{\mathbf{x}}_0) = \mathbf{0} \quad .$$

It follows that $\mathbf{K}$ must be singular (i.e. $\det(\mathbf{K}) = 0$), which means that at least one of its eigenvalues, say λ_1, must vanish: $\lambda_1 = 0$. If we represent $\dot{\mathbf{x}}$ by means of the right eigenvectors of $\mathbf{K}$ we obtain

$$\mathbf{K}(\mathbf{w}_i \cdot \dot{\mathbf{x}})\mathbf{v}_i = \lambda_i(\mathbf{w}_i \cdot \dot{\mathbf{x}})\mathbf{v}_i = \dot{\mathbf{y}} \qquad\qquad i = 2, 3, \ldots n$$

with $\lambda_2 \cdot \lambda_3 \ldots \lambda_n \neq 0$. Hence

$$\lambda_i[\mathbf{w}_i \cdot (\dot{\mathbf{x}}_1 - \dot{\mathbf{x}}_0)]\mathbf{v}_1 = \mathbf{0} \quad .$$

It then follows $\mathbf{w}_i \cdot (\dot{\mathbf{x}}_1 - \dot{\mathbf{x}}_0) = 0$ for $i = 2, 3, \ldots n$. This means that $\dot{\mathbf{x}}_1 - \dot{\mathbf{x}}_0$ has the direction of $\mathbf{v}_1$:

$$\dot{\mathbf{x}}_1 - \dot{\mathbf{x}}_0 = \alpha\mathbf{v}_1$$

$$\dot{\mathbf{x}}_1 = \dot{\mathbf{x}}_0 + \alpha\mathbf{v}_1 \quad . \tag{5.21}$$

For any $\alpha = 0$ we obtain with (5.21) a bifurcated solution. It is reported [70] that localized solutions are often orthogonal to the homogeneous deformation solution $\dot{\mathbf{x}}_0$, i.e. $\dot{\mathbf{x}}_1 \cdot \dot{\mathbf{x}}_0 = 0$. This can be obtained with $\alpha = -(\dot{\mathbf{x}}_0 \cdot \mathbf{v}_1)/(\mathbf{v}_1 \cdot \mathbf{v}_1)$. Some numerical methods to find the eigenvectors of $\mathbf{K}$ for zero (or even negative) eigenvalues are discussed in [70]. To trace solutions of the type (5.21) is called 'eigenvector perturbation'.

Another method which circumvents the search for eigenvectors is the so-called material perturbation (see [53]): The material properties are assumed to scatter over their mean values. Then, the solution of (5.20) traces automatically the bifurcated solution. This is probably due to the onset of ill-posedness, according to which small perturbations grow exponentially.

For hypoplastic materials (and also for elastoplastic materials, a fact which is often overlooked) the stiffness matrix $\mathbf{K}_{ij}$ is not constant, therefore the eigenvector perturbation makes no sense. Material perturbation is however still applicable. Another possible procedure is to search the vector $\mathbf{x}$ that minimizes the second order work

$$f(\mathbf{x}) = L_{ij}\dot{x}_i\dot{x}_j + N_{ip}\sqrt{a_{pxl}\ddot{x}_k\dot{x}_l}\,\dot{x}_i$$

and then to check whether this minimum is equal to zero.

Chapter 6

Exercises

For the exercises stated below use the hypoplastic constitutive equation version 4.3 with the following material constants:

$$C_1 = -106.5, \quad C_2 = -801.5, \quad C_3 = -797.1, \quad C_4 = 1077.7 \quad .$$

1. Given $\mathbf{T} = \begin{pmatrix} -20\,\text{kN/m}^2 & 0 & 0 \\ 0 & -10\,\text{kN/m}^2 & 0 \\ 0 & 0 & -10\,\text{kN/m}^2 \end{pmatrix}$

   and $\mathbf{D} = \begin{pmatrix} -1s^{-1} & 0 & 0 \\ 0 & 0 & 0 \\ 0 & 0 & 0 \end{pmatrix}$. Determine $\overset{\circ}{\mathbf{T}}$.

2. Given $\mathbf{T}$ as above and $\mathbf{D} = \begin{pmatrix} -1s^{-1} & 0 & 0 \\ 0 & D_2 & 0 \\ 0 & 0 & D_2 \end{pmatrix}$.

   Determine D_2 in such a way that $\overset{\circ}{T}_2 = \overset{\circ}{T}_3 = 0$.
   Hint: An algebraic equation has to be solved (e.g. by trial and error).

3. Referring to the results shown in fig. 2.3 determine the tensors $\mathbf{T}, \mathbf{D}$ and $\overset{\circ}{\mathbf{T}}$ for $\sigma_2 = 100$ kPa and $\varepsilon_1 = -4\%$.

4. Numerical simulation of an oedometric test: Starting from the stress state $\sigma_1 = \sigma_2 = \sigma_3 = 100$ kN/m^2 the stress component σ_1 is increased up to 1000 kN/m^2 and then reduced to the original value, i.e. $\sigma_1 = 100 \ldots 1000 \ldots 100$ kN/m^2. No lateral deformation is allowed for, i.e. $D_2 \equiv D_3 \equiv 0$. Plot the σ_1 vs. σ_2 curve (stress path) and the $\log \sigma_1$ vs. ε_1 curve (stress-strain-curve).

5. Numerical simulation of a triaxial compression test: Starting from the stress state $\sigma_1 = \sigma_2 = \sigma_3 = 100$ kN/m^2 the axial strain $|\varepsilon_1|$ is increased from 0 to 4% keeping $\sigma_2 = \sigma_3$ constant. Subsequently σ_1 is reduced by unloading to its initial value. Plot the stress-strain-curve $(\sigma_1 - \sigma_2)$ vs. $|\varepsilon_1|$ and the volumetric curve ε_v vs. $|\varepsilon_1|$.

66

6. Numerical simulation of a so-called deviatoric test: Starting from the stress state $\sigma_1 = \sigma_2 = \sigma_3 = 100$ kN/m^2, compression is increased from zero to $|\varepsilon_1| = 4\%$ keeping the sum $\sigma_1 + \sigma_2 + \sigma_3$ constant. Subsequently σ_1 is reduced by unloading to its initial value. Plot the stress-strain-curve $(\sigma_1 - \sigma_2)$ vs. $|\varepsilon_1|$ and the volumetric curve ε_v vs. $|\varepsilon_1|$.

7. Numerical simulation of undrained triaxial compression: Starting from the stress state $\sigma_1' = \sigma_2' = \sigma_3' = 100$ kN/m^2 and keeping $\sigma_2' = \sigma_3' = 100$ kN/m^2 constant, a water-saturated sample is compressed with closed drainage from zero to $|\varepsilon_1| = 4\%$. Plot the effective stress-path σ_1' vs. σ_2', the stress strain curve $(\sigma_1' - \sigma_2')$ vs. $|\varepsilon_1|$ and the pore pressure u vs. $|\varepsilon_1|$. The initial pore pressure is zero.

8. Plot the response envelopes for the stress states

   (i) $\sigma_1 = \sigma_2 = \sigma_3 = 100$ kN/m^2,

   (ii) $\sigma_1 = 200$ kN/m^2; $\sigma_2 = \sigma_3 = 100$ kN/m^2.

9. Given the stress state $\mathbf{T}$ with $\sigma_1 = 150$ kN/m^2; $\sigma_2 = \sigma_3 = 100$ kN/m^2. Determine $\mathbf{D}$ such that $\overset{\circ}{\mathbf{T}} = \lambda\mathbf{T}$.

10. For a given $\mathbf{D}$-tensor, say $\mathbf{D} = \begin{pmatrix} -1 & 0 & 0 \\ 0 & -2 & 0 \\ 0 & 0 & 0 \end{pmatrix}$, determine $\mathbf{T}$ such that

    $\overset{\circ}{\mathbf{T}} = \lambda\mathbf{T}$.

11. Given the stress state $\mathbf{T}$ with $\sigma_1 = 150$ kN/m^2; $\sigma_2 = \sigma_3 = 100$ kN/m^2. Plot the stress paths (σ_1 vs. σ_2 curves) emanating from $\mathbf{T}$ and obtained with $\mathbf{D} = $ const. $\mathbf{D}$ should be taken as

    $$D = \begin{pmatrix} -\cos\alpha & 0 & 0 \\ 0 & -\sin\alpha/\sqrt{2} & 0 \\ 0 & 0 & -\sin\alpha/\sqrt{2} \end{pmatrix}$$

    with $\alpha = 0°, 10°, 20°, \ldots 340°, 350°$.

12. Plot the cross-section of the limit surface with the deviatoric plane $\sigma_1 + \sigma_2 + \sigma_3 = 300$ kN/m^2.

Solutions:

1. With tr $\mathbf{T} = -20 - 10 - 10 = -40$ kN/m^2,

   $$\mathbf{T}^* = \begin{pmatrix} -20 & 0 & 0 \\ 0 & -10 & 0 \\ 0 & 0 & -10 \end{pmatrix} - \frac{-40}{3}\begin{pmatrix} 1 & 0 & 0 \\ 0 & 1 & 0 \\ 0 & 0 & 1 \end{pmatrix} = \begin{pmatrix} -6.66 & 0 & 0 \\ 0 & 3.33 & 0 \\ 0 & 0 & 3.33 \end{pmatrix},$$

$$\mathbf{T}^{*2} = \begin{pmatrix} 44.44 & 0 & 0 \\ 0 & 11.11 & 0 \\ 0 & 0 & 11.11 \end{pmatrix} ,$$

$$\mathbf{T}^2 = \begin{pmatrix} 400 & 0 & 0 \\ 0 & 100 & 0 \\ 0 & 0 & 100 \end{pmatrix} , \quad \mathbf{D}^2 = \begin{pmatrix} 1 & 0 & 0 \\ 0 & 0 & 0 \\ 0 & 0 & 0 \end{pmatrix} ,$$

$$\operatorname{tr} \mathbf{D}^2 = 1 \ , \quad \sqrt{\operatorname{tr} \mathbf{D}^2} = 1 \ ,$$

$$\mathbf{TD} = \begin{pmatrix} 20 & 0 & 0 \\ 0 & 0 & 0 \\ 0 & 0 & 0 \end{pmatrix} , \quad \operatorname{tr}(\mathbf{TD}) = 20 \ ,$$

the stress rate $\mathring{\mathbf{T}}$ is composed of

$$C_1 \cdot \operatorname{tr} \mathbf{T} \cdot \mathbf{D} = (-106.5) \cdot (-40) \begin{pmatrix} -1 & 0 & 0 \\ 0 & 0 & 0 \\ 0 & 0 & 0 \end{pmatrix} = \begin{pmatrix} -4260 & 0 & 0 \\ 0 & 0 & 0 \\ 0 & 0 & 0 \end{pmatrix}$$

$$C_2 \frac{\operatorname{tr}(\mathbf{TD})}{\operatorname{tr} \mathbf{T}} \mathbf{T} = (-801.5) \cdot \frac{20}{-40} \begin{pmatrix} -20 & 0 & 0 \\ 0 & -10 & 0 \\ 0 & 0 & -10 \end{pmatrix}$$

$$= \begin{pmatrix} -8015 & 0 & 0 \\ 0 & -4007.5 & 0 \\ 0 & 0 & -4007.5 \end{pmatrix}$$

$$C_3 \frac{\mathbf{T}^2}{\operatorname{tr} \mathbf{T}} \sqrt{\operatorname{tr} \mathbf{D}^2} = -797.1 \cdot \frac{1}{-40} \begin{pmatrix} 400 & 0 & 0 \\ 0 & 100 & 0 \\ 0 & 0 & 100 \end{pmatrix} \cdot 1$$

$$= \begin{pmatrix} 7971 & 0 & 0 \\ 0 & 1993.75 & 0 \\ 0 & 0 & 1993.75 \end{pmatrix}$$

$$C_4 \frac{\mathbf{T}^{*2}}{\operatorname{tr} \mathbf{T}} \sqrt{\operatorname{tr} \mathbf{D}^2} = 1077.7 \cdot \frac{1}{40} \cdot \begin{pmatrix} 44.44 & 0 & 0 \\ 0 & 11.11 & 0 \\ 0 & 0 & 11.11 \end{pmatrix}$$

$$= \begin{pmatrix} -1197.44 & 0 & 0 \\ 0 & -299.361 & 0 \\ 0 & 0 & -299.361 \end{pmatrix} ,$$

$$\text{hence} \quad \mathring{\mathbf{T}} = \begin{pmatrix} -5501.4 & 0 & 0 \\ 0 & -2313.1 & 0 \\ 0 & 0 & -2313.1 \end{pmatrix} .$$

2. $D_2 = D_3$ and $T_2 = T_3$ imply axisymmetric conditions and, thus, $\mathring{\mathbf{T}}_2 = \mathring{\mathbf{T}}_3$.

With

$$\operatorname{tr}\mathbf{T} = T_1 + 2T_2 \quad,$$

$$\mathbf{TD} = \begin{pmatrix} T_1 D_1 & 0 & 0 \\ 0 & T_2 D_2 & 0 \\ 0 & 0 & T_2 D_2 \end{pmatrix} \quad,$$

$$\operatorname{tr}(\mathbf{TD}) = T_1 D_1 + 2T_2 D_2 \quad,$$

$$\mathbf{D}^2 = \begin{pmatrix} D_1^2 & 0 & 0 \\ 0 & D_2^2 & 0 \\ 0 & 0 & D_2^2 \end{pmatrix} \quad,$$

$$\sqrt{\operatorname{tr}\mathbf{D}^2} = \sqrt{D_1^2 + 2D_2^2} \quad,$$

$$\mathbf{T}^2 = \begin{pmatrix} T_1^2 & 0 & 0 \\ 0 & T_2^2 & 0 \\ 0 & 0 & T_2^2 \end{pmatrix} \quad,$$

$$\mathbf{T}^* = \begin{pmatrix} T_1 & 0 & 0 \\ 0 & T_2 & 0 \\ 0 & 0 & T_2 \end{pmatrix} - \frac{T_1 + 2T_2}{3} \begin{pmatrix} 1 & 0 & 0 \\ 0 & 1 & 0 \\ 0 & 0 & 1 \end{pmatrix}$$

$$= \frac{1}{3} \begin{pmatrix} 2(T_1 - T_2) & 0 & 0 \\ 0 & T_2 - T_1 & 0 \\ 0 & 0 & T_2 - T_1 \end{pmatrix}$$

$$\mathbf{T}^{*2} = \frac{1}{9} \begin{pmatrix} 4(T_1 - T_2)^2 & 0 & 0 \\ 0 & (T_1 - T_2)^2 & 0 \\ 0 & 0 & (T_1 - T_2)^2 \end{pmatrix}$$

we obtain

$$\dot{T}_2 = C_1(T_1 + 2T_2)D_2 + C_2 \frac{T_1 D_1 + 2T_2 D_2}{T_1 + +2T_2} T_2 + C_3 \frac{T_2^2 \sqrt{D_1^2 + 2D_2^2}}{T_1 + 2T_2}$$
$$+ \; C_4 \frac{\frac{1}{9}(T_1 - T_2)^2 \sqrt{D_1^2 + 2D_2^2}}{T_1 + 2T_2} \quad.$$

Introducing the values for T_1, T_2, D_1 we obtain the following equation expressing $\dot{T}_2 = 0$:

$$-4007.5 + 8267.5 D_2 + 1693.38\sqrt{1 + 2D_2^2} = 0$$

with the solution $D_2 = 0.2659$.

3. From $T_1/T_2 = 5.3$ follows $T_1 = 5.3 \cdot (-100) = -530$ kN/m^2, thus

$$T = \begin{pmatrix} -530 & 0 & 0 \\ 0 & -100 & 0 \\ 0 & 0 & -100 \end{pmatrix} \quad .$$

From equ. (4.5) we infer

$$\tan \beta = 1 + 2\frac{\dot\varepsilon_2}{\dot\varepsilon_1} \quad .$$

With $\tan \beta = -0.65$ (from fig. 2.3) and $\dot\varepsilon_1 = -1$ (the value of $|\varepsilon_1|$ is immaterial, as sand is considered as rate independent) we obtain $\varepsilon_2 = 0.825$, i.e.

$$D = \begin{pmatrix} -1 & 0 & 0 \\ 0 & 0.825 & 0 \\ 0 & 0 & 0.825 \end{pmatrix} \quad .$$

The slope of the σ_1/σ_2 vs. $|\varepsilon_1|$ curve for $\varepsilon_1 = 4\%$ is

$$\frac{\Delta(\sigma_1/\sigma_2)}{\Delta(-\varepsilon_1)} = 1.4$$

i.e.

$$\Delta\sigma_1 = 1.4\sigma_2\Delta(-\varepsilon_1) \quad \text{or} \quad \Delta\sigma_1/\Delta t = -1.4\sigma_2\Delta\varepsilon_1/\Delta t$$

$$\dot\sigma_1 = -1.4\sigma_2\dot\varepsilon_1$$

With $\dot\varepsilon_1 = -1$ we obtain $\dot\sigma_1 = -140$, thus

$$\overset{..}{T} = \begin{pmatrix} -140 & 0 & 0 \\ 0 & 0 & 0 \\ 0 & 0 & 0 \end{pmatrix} \quad \frac{\text{kN}}{\text{m}^2\text{s}}$$

4. Oedometric test: given stress state: $\sigma_1 = \sigma_2 = \sigma_3 = 100$ kN/m^2 $\rightarrow T_1 = T_2 = T_3 = -100$ kN/m^2. The kinematical boundary conditions of an oedometric test ($D_2 = D_3 = 0$) imply axial symmetry with $T_2 = T_3$ and $\dot T_2 = \dot T_3$.

$$D = \begin{pmatrix} D_1 & 0 & 0 \\ 0 & 0 & 0 \\ 0 & 0 & 0 \end{pmatrix}, \quad T = \begin{pmatrix} T_1 & 0 & 0 \\ 0 & T_2 & 0 \\ 0 & 0 & T_2 \end{pmatrix},$$

$$TD = \begin{pmatrix} T_1 D_1 & 0 & 0 \\ 0 & 0 & 0 \\ 0 & 0 & 0 \end{pmatrix},$$

$$\text{tr}\,T = T_1 + 2T_2, \quad \text{tr}(TD) = T_1 D_1,$$

$$\mathbf{D}^2 = \begin{pmatrix} D_1^2 & 0 & 0 \\ 0 & 0 & 0 \\ 0 & 0 & 0 \end{pmatrix} \quad ,$$

$$\text{tr}\mathbf{D}^2 = D_1^2$$

$$\sqrt{\text{tr}\mathbf{D}^2} = \sqrt{D_1^2} = |D_1| = \begin{cases} +D_1 & \text{for} \quad D_1 > 0 \quad \text{(unloading)} \\ -D_1 & \text{for} \quad D_1 < 0 \quad \text{(loading)} \end{cases} \quad ,$$

$$\mathbf{T}^2 = \begin{pmatrix} T_1^2 & 0 & 0 \\ 0 & T_2^2 & 0 \\ 0 & 0 & T_3^2 \end{pmatrix} \quad ,$$

$$\mathbf{T}^* = \begin{pmatrix} T_1 & 0 & 0 \\ 0 & T_2 & 0 \\ 0 & 0 & T_2 \end{pmatrix} - \frac{T_1 + 2T_2}{3} \begin{pmatrix} 1 & 0 & 0 \\ 0 & 1 & 0 \\ 0 & 0 & 1 \end{pmatrix}$$

$$= \frac{1}{3} \begin{pmatrix} 2(T_1 - T_2) & 0 & 0 \\ 0 & (T_2 - T_1) & 0 \\ 0 & 0 & (T_2 - T_1) \end{pmatrix} \quad ,$$

$$\mathbf{T}^{*2} = \frac{1}{9} \begin{pmatrix} 4(T_1 - T_2)^2 & 0 & 0 \\ 0 & (T_2 - T_1)^2 & 0 \\ 0 & 0 & (T_2 - T_1^2) \end{pmatrix} \quad ,$$

$$\mathring{\mathbf{T}} = \begin{pmatrix} \dot{T}_1 & 0 & 0 \\ 0 & \dot{T}_2 & 0 \\ 0 & 0 & \dot{T}_3 \end{pmatrix}$$

$$= C_1(T_1 + 2T_2) \begin{pmatrix} D_1 & 0 & 0 \\ 0 & 0 & 0 \\ 0 & 0 & 0 \end{pmatrix} + C_2 \frac{T_1 D_1}{T_1 + 2T_2} \begin{pmatrix} T_1 & 0 & 0 \\ 0 & T_2 & 0 \\ 0 & 0 & T_3 \end{pmatrix}$$

$$+ C_3 \begin{pmatrix} T_1^2 & 0 & 0 \\ 0 & T_2^2 & 0 \\ 0 & 0 & T_3^2 \end{pmatrix} \frac{|D_1|}{T_1 + 2T_2} \tag{6.1}$$

$$+ C_4 \frac{1}{9} \begin{pmatrix} 4(T_1 - T_2)^2 & 0 & 0 \\ 0 & (T_1 - T_2)^2 & 0 \\ 0 & 0 & (T_1 - T_2)^2 \end{pmatrix} \frac{|D_1|}{T_1 + 2T_2}$$

The matrix equation (6.1) contains 9 scalar equations (one for each matrix component). However, only two of them are non-trivial:

$$\dot{T}_1 = C_1(T_1 + 2T_2)D_1 + C_2 \frac{T_1 D_1}{T_1 + 2T_2} T_1$$

$$+ \quad C_3 \frac{T_1^2 |D_1|}{T_1 + 2T_2} + C_4 \frac{4}{9}(T_1 - T_2)^2 \frac{|D_1|}{T_1 + 2T_2}$$

$$\dot{T}_2 \quad = \quad C_2 \frac{T_1 D_1}{T_1 + 2T_2} T_2 \qquad\qquad\qquad (6.2)$$

$$+ \quad C_3 \frac{T_2^2 |D_1|}{T_1 + 2T_2} + C_4 \frac{1}{9}(T_1 - T_2)^2 \frac{|D_1|}{T_1 + 2T_2}$$

We can put $D_1 = -1(\mathrm{s}^{-1})$ for compression (loading), and $D_1 = 1(\mathrm{s}^{-1})$ for unloading. Due to rate independence, the value of $|D_1|$ is immaterial. Multiplying both sides of (6.2) with Δt we can set $\dot{T}_1 \Delta t \approx \Delta T_1$, $\dot{T}_2 \Delta t \approx \Delta T_2$, $D_1 \Delta t \approx \Delta \varepsilon_1$. The smaller Δt is, the preciser these equations are. The numerical integration (so-called EULER-forward integration) proceeds along the scheme:

$$T_1^{(1)} = T_1^{(0)} + \quad C_1\left(T_1^{(0)} + 2T_2^{(0)}\right) D_1 \Delta t \quad +C_2 \frac{T_1^{(0)} D_1}{T_1^{(0)} + 2T_2^{(0)}} T_1^{(0)} \Delta t + C_3 \cdots$$

$$T_2^{(1)} = T_2^{(0)} \qquad\qquad\qquad\qquad\qquad +C_2 \frac{T_1^{(0)} D_1}{T_1^{(0)} + 2T_2^{(0)}} T_2^{(0)} \Delta t + C_3 \cdots$$

$$\varepsilon_1^{(1)} = \qquad\qquad \varepsilon_1^{(0)} + D_1 \Delta t$$

$$T_1^{(2)} = T_1^{(1)} + \quad C_1\left(T_1^{(1)} + 2T_2^{(1)}\right) D_1 \Delta t \quad +C_2 \frac{T_1^{(1)} D_1}{T_1^{(1)} + 2T_2^{(1)}} T_1^{(1)} \Delta t + C_3 \cdots$$

$$T_2^{(2)} = T_2^{(1)} \qquad\qquad\qquad\qquad\qquad +C_2 \frac{T_1^{(1)} D_1}{T_1^{(1)} + 2T_2^{(1)}} T_2^{(1)} \Delta t + C_3 \cdots$$

$$\varepsilon_1^{(2)} = \qquad\qquad \varepsilon_1^{(1)} + D_1 \Delta t$$

$$\cdots (\text{see table 6.1})$$

As result we obtain numerically the curves $\sigma_1(\varepsilon_1)$, $\sigma_2(\varepsilon_1)$ (see fig.6.1). The oedometric condition implies $\dot{\varepsilon}_2 = 0$. Δt should be sufficiently small. To obtain this we start with a chosen Δt, which we consider to be small enough. We carry out the numerical integration and then we repeat it with, say, $\Delta t := \Delta t / 10$. If we obtain identical $\sigma_1(\varepsilon_1)$ and $\sigma_2(\varepsilon_1)$ curves, then the initially chosen Δt was indeed small enough.

5. Triaxial compression test:

With $\mathbf{D} = \begin{pmatrix} D_1 & 0 & 0 \\ 0 & D_2 & 0 \\ 0 & 0 & D_2 \end{pmatrix}$ and $\mathbf{T} = \begin{pmatrix} T_1 & 0 & 0 \\ 0 & T_2 & 0 \\ 0 & 0 & T_2 \end{pmatrix}$ the constitutive equation for the stress-rate $\dot{T}_1$ reads

$$\dot{T}_1 \quad = \quad C_1(T_1 + 2T_2)D_1 + C_2 \frac{T_1 D_1 + 2T_2 D_2}{T_1 + 2T_2} T_1 + C_3 \frac{T_1^2 \sqrt{D_1^2 + 2D_2^2}}{T_1 + 2T_2}$$

Table 6.1: Numerical simulation of an oedometric test

step	Δt [s]	D_1 [1/s]	ε_1 [%]	T_1 [kN/m^2]	T_2 [kN/m^2]
0		1	0.00	−100.0	−100.0
1	0.001	−1	−0.10	−132.1	−100.1
2	0.001	−1	−0.20	−169.2	−108.0
3	0.001	−1	−0.30	−215.2	−121.9
4	0.001	−1	−0.40	−273.6	−141.9
5	0.001	−1	−0.50	−348.5	−169.0
6	0.001	−1	−0.60	−444.8	−204.6
7	0.001	−1	−0.70	−569.2	−251.0
8	0.001	−1	−0.80	−729.9	−311.1
9	0.001	−1	−0.90	−937.7	−388.6
10	0.001	−1	−1.00	−1206.8	−488.8
11	0.0002	1	−0.98	−969.8	−428.1
12	0.0002	1	−0.96	−781.6	−375.6
13	0.0002	1	−0.94	−631.8	−330.3
14	0.0002	1	−0.92	−512.3	−290.9
15	0.0002	1	−0.90	−416.6	−256.8
16	0.0002	1	−0.88	−339.7	−227.0
17	0.0002	1	−0.86	−277.9	−201.1
18	0.0002	1	−0.84	−227.9	−178.5
19	0.0002	1	−0.82	−187.5	−158.6
20	0.0002	1	−0.80	−154.6	−141.2
21	0.0002	1	−0.78	−127.9	−126.0

$$+ \quad C_4 \frac{4}{9} (T_1 - T_2)^2 \frac{\sqrt{D_1^2 + 2D_2^2}}{T_1 + 2T_2}$$

This equation can be numerically integrated (in a similar way as stated in exercise 4) to obtain the curves $(\sigma_1 - \sigma_2)$ vs. ε_1, and ε_v vs. ε_1 (see fig.6.2). We set $D_1 = -1$ for loading and $D_1 = 1$ for unloading. Note that D_2 is unknown. It has, therefore, to be determined at each integration step by solving the equation (cf. exercise 2):

$$\dot{T}_2 = C_1(T_1 + 2T_2)D_2 + C_2 \frac{T_1 D_1 + 2T_2 D_2}{T_1 + 2T_2} T_2 + C_3 \frac{T_2^2 \sqrt{D_1^2 + 2D_2^2}}{T_1 + 2T_2}$$

$$+ \quad C_4 \frac{1}{9} (T_1 - T_2)^2 \frac{\sqrt{D_1^2 + 2D_2^2}}{T_1 + 2T_2} = 0,$$

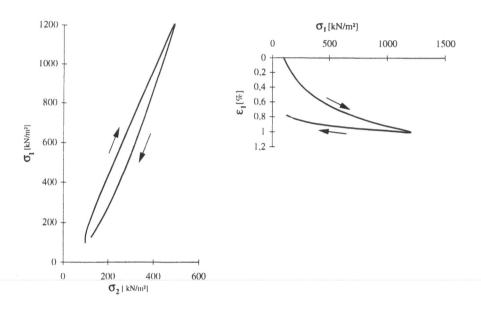

Figure 6.1: Numerical simulation of an oedometric test

which expresses the static boundary conditions ($\sigma_2 = \sigma_3 = $ const, $\leadsto \dot{\sigma}_2 = \dot{\sigma}_3 = 0$, i.e. $\dot{T}_2 = \dot{T}_3 = 0$). Due to axisymmetic conditions the constitutive equation correponds to only two non-trivial scalar equations. The result of the integration (using $\Delta t = 0.001$ for loading and $\Delta t = 0.0001$ for unloading) is shown in table 6.2.

6. Triaxial deviatoric test:

We have again $\mathbf{D} = \begin{pmatrix} D_1 & 0 & 0 \\ 0 & D_2 & 0 \\ 0 & 0 & D_3 \end{pmatrix}$, $\mathbf{T} = \begin{pmatrix} T_1 & 0 & 0 \\ 0 & T_2 & 0 \\ 0 & 0 & T_3 \end{pmatrix}$ with

axisymmetric conditions ($D_2 = D_3$ and $T_2 = T_3$) and proceed as in exercise 5. The only difference to the previous exercise is that D_2 has now to be determined so as to fulfil the static boundary condition tr$\mathbf{T}$ = const (or $\sigma_1 + 2\sigma_2 = $ const, wich means $\dot{\sigma}_1 + 2\dot{\sigma}_2 = 0$). I.e., D_2 has to be determined at each integration step by solving the equation $\dot{T}_1 + 2\dot{T}_2 = 0$:

$$
\begin{aligned}
\dot{T}_1 + 2\dot{T}_2 &= C_1(T_1 + 2T_2)(D_1 + 2D_2) + C_2 \frac{T_1 D_1 + 2T_2 D_2}{T_1 + 2T_2}(T_1 + 2T_2) \\
&+ C_3(T_1^2 + 2T_2^2)\frac{\sqrt{D_1^2 + 2D_2^2}}{T_1 + 2T_2} \\
&+ C_4 \frac{1}{9}\left(4(T_1 - T_2)^2 + 2(T_2 - T_1)^2\right)\frac{\sqrt{D_1^2 + 2D_2^2}}{T_1 + 2T_2} = 0,
\end{aligned}
$$

Table 6.2: Numerical simulation of a triaxial compression test

step	Δt [s]	D_1 [s^{-1}]	D_2 [s^{-1}]	ε_1 [%]	$\varepsilon_2 = \varepsilon_3$ [%]	ε_v [%]	σ_1 [kN/m^2]	$\sigma_2 = \sigma_3$ [kN/m^2]
0				0.00	0.00	0.00	100.0	100.0
1	0.001	−1	0.00	−0.10	0.00	−0.10	132.0	100.0
2	0.001	−1	0.10	−0.20	0.01	−0.18	162.6	100.0
3	0.001	−1	0.18	−0.30	0.03	−0.25	192.4	100.0
4	0.001	−1	0.25	−0.40	0.05	−0.30	221.6	100.0
5	0.001	−1	0.31	−0.50	0.08	−0.33	250.2	100.0
⋮	⋮	⋮	⋮	⋮	⋮	⋮	⋮	⋮
10	0.001	−1	0.57	−1.00	0.32	−0.35	373.6	100.0
⋮	⋮	⋮	⋮	⋮	⋮	⋮	⋮	⋮
15	0.001	−1	0.72	−1.50	0.66	−0.18	440.8	100.0
⋮	⋮	⋮	⋮	⋮	⋮	⋮	⋮	⋮
20	0.001	−1	0.77	−2.00	1.04	0.07	463.2	100.0
⋮	⋮	⋮	⋮	⋮	⋮	⋮	⋮	⋮
25	0.001	−1	0.78	−2.50	1.43	0.35	468.9	100.0
⋮	⋮	⋮	⋮	⋮	⋮	⋮	⋮	⋮
30	0.001	−1	0.79	−3.00	1.82	0.64	470.2	100.0
⋮	⋮	⋮	⋮	⋮	⋮	⋮	⋮	⋮
35	0.001	−1	0.79	−3.50	2.21	0.93	470.6	100.0
36	0.001	−1	0.79	−3.60	2.29	0.98	470.6	100.0
37	0.001	−1	0.79	−3.70	2.37	1.04	470.6	100.0
38	0.001	−1	0.79	−3.80	2.45	1.10	470.6	100.0
39	0.001	−1	0.79	−3.90	2.53	1.16	470.6	100.0
40	0.001	−1	0.79	−4.00	2.61	1.21	470.6	100.0
41	0.0001	1	−0.43	−3.99	2.60	1.22	422.5	100.0
42	0.0001	1	−0.50	−3.98	2.60	1.22	380.2	100.0
43	0.0001	1	−0.55	−3.97	2.59	1.21	342.8	100.0
44	0.0001	1	−0.60	−3.96	2.59	1.21	310.0	100.0
45	0.0001	1	−0.65	−3.95	2.58	1.21	281.0	100.0
⋮	⋮	⋮	⋮	⋮	⋮	⋮	⋮	⋮
50	0.0001	1	−0.81	−3.90	2.54	1.18	179.9	100.0
⋮	⋮	⋮	⋮	⋮	⋮	⋮	⋮	⋮
55	0.0001	1	−0.81	−3.85	2.50	1.15	124.1	100.0
56	0.0001	1	−0.81	−3.84	2.49	1.15	116.2	100.0
57	0.0001	1	−0.80	−3.83	2.49	1.14	109.0	100.0
58	0.0001	1	−0.79	−3.82	2.48	1.14	102.5	100.0
59	0.0001	1	−0.78	−3.81	2.47	1.13	96.7	100.0

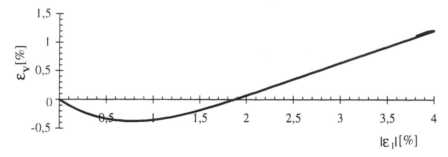

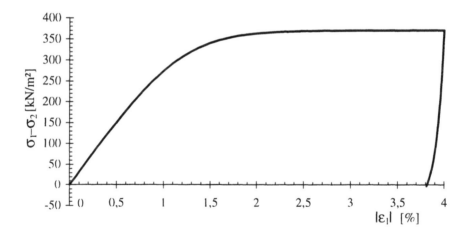

Figure 6.2: Numerical simulation of a triaxial compression test

The numerical integration (using $\Delta t = 0.001$ for loading and $\Delta t = 0.00005$ for unloading) yields the results shown in table 6.3 and plotted in fig. 6.3.

Table 6.3: Numerical simulation of a deviatoric test

step	Δt [s]	D_1 [s^{-1}]	D_2 [s^{-1}]	ε_1 [%]	$\varepsilon_2 = \varepsilon_3$ [%]	ε_v [%]	σ_1 [kN/m^2]	$\sigma_2 = \sigma_3$ [kN/m^2]
0				0.00	0.00	0.00	100.0	100.0
1	0.001	-1	0.14	-0.10	0.01	-0.07	124.2	87.9
2	0.001	-1	0.24	-0.20	0.04	-0.13	142.6	78.7
3	0.001	-1	0.32	-0.30	0.07	-0.16	157.3	71.4
4	0.001	-1	0.40	-0.40	0.11	-0.18	169.2	65.4
5	0.001	-1	0.47	-0.50	0.16	-0.19	178.9	60.6
$\vdots$	$\vdots$	$\vdots$	$\vdots$	$\vdots$	$\vdots$	$\vdots$	$\vdots$	$\vdots$
10	0.001	-1	0.71	-1.00	0.47	-0.06	203.9	48.0
$\vdots$	$\vdots$	$\vdots$	$\vdots$	$\vdots$	$\vdots$	$\vdots$	$\vdots$	$\vdots$
15	0.001	-1	0.77	-1.50	0.85	0.20	209.4	45.3
$\vdots$	$\vdots$	$\vdots$	$\vdots$	$\vdots$	$\vdots$	$\vdots$	$\vdots$	$\vdots$
20	0.001	-1	0.79	-2.00	1.24	0.48	210.4	44.8
$\vdots$	$\vdots$	$\vdots$	$\vdots$	$\vdots$	$\vdots$	$\vdots$	$\vdots$	$\vdots$
25	0.001	-1	0.79	-2.50	1.63	0.77	210.5	44.7
$\vdots$	$\vdots$	$\vdots$	$\vdots$	$\vdots$	$\vdots$	$\vdots$	$\vdots$	$\vdots$
30	0.001	-1	0.79	-3.00	2.03	1.06	210.5	44.7
$\vdots$	$\vdots$	$\vdots$	$\vdots$	$\vdots$	$\vdots$	$\vdots$	$\vdots$	$\vdots$
35	0.001	-1	0.79	-3.50	2.42	1.35	210.5	44.7
36	0.001	-1	0.79	-3.60	2.50	1.40	210.5	44.7
37	0.001	-1	0.79	-3.70	2.58	1.46	210.5	44.7
38	0.001	-1	0.79	-3.80	2.66	1.52	210.5	44.7
39	0.001	-1	0.79	-3.90	2.74	1.58	210.5	44.7
40	0.001	-1	0.79	-4.00	2.82	1.64	210.5	44.7
41	0.00005	1	-4.37	-4.00	2.80	1.60	190.9	54.6
42	0.00005	1	-3.49	-3.99	2.78	1.57	177.2	61.4
43	0.00005	1	-2.97	-3.99	2.76	1.54	166.7	66.7
44	0.00005	1	-2.62	-3.98	2.75	1.52	158.1	71.0
45	0.00005	1	-2.37	-3.98	2.74	1.50	150.7	74.6
$\vdots$	$\vdots$	$\vdots$	$\vdots$	$\vdots$	$\vdots$	$\vdots$	$\vdots$	$\vdots$
50	0.00005	1	-1.68	-3.95	2.69	1.43	124.9	87.5
$\vdots$	$\vdots$	$\vdots$	$\vdots$	$\vdots$	$\vdots$	$\vdots$	$\vdots$	$\vdots$
55	0.00005	1	-1.36	-3.93	2.65	1.38	108.1	95.9
56	0.00005	1	-1.32	-3.92	2.65	1.37	105.3	97.3
57	0.00005	1	-1.28	-3.92	2.64	1.37	102.7	98.6
58	0.00005	1	-1.24	-3.91	2.63	1.36	100.2	99.9
59	0.00005	1	-1.20	3.91	2.63	1.35	97.9	101.1

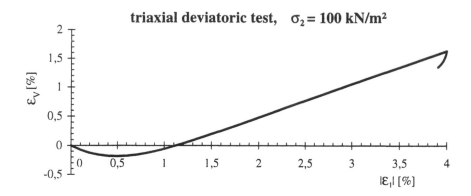

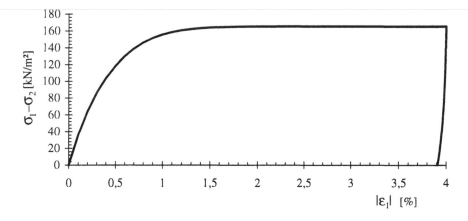

Figure 6.3: Numerical simulation of a deviatoric test

7. To solve this exercise we must take into account that all stresses appearing in the hypoplastic constitutive equation are, actually, effective stresses such that the constitutive equation should be rewritten as $\overset{\circ}{\mathbf{T}}{}' = \mathbf{h}(\mathbf{T}', \mathbf{D})$. Considering the pore pressure u as positive at compression we have the relations

$$\sigma_{ij} = \sigma'_{ij} + u\delta_{ij}, \quad \text{or} \quad -T_{ij} = -T'_{ij} + u\delta_{ij}$$

i.e.

$$\begin{pmatrix} \sigma_{11} & \sigma_{12} & \sigma_{13} \\ \sigma_{21} & \sigma_{22} & \sigma_{23} \\ \sigma_{31} & \sigma_{32} & \sigma_{33} \end{pmatrix} = \begin{pmatrix} \sigma'_{11} & \sigma'_{12} & \sigma'_{13} \\ \sigma'_{21} & \sigma'_{22} & \sigma'_{23} \\ \sigma'_{31} & \sigma'_{32} & \sigma'_{33} \end{pmatrix} + \begin{pmatrix} u & 0 & 0 \\ 0 & u & 0 \\ 0 & 0 & u \end{pmatrix}.$$

The undrained condition (i.e. condition of constant volume) can be expressed as $\dot{\varepsilon}_v = \text{tr}\mathbf{D} = 0$, or $D_1 + 2D_2 = 0$, or $D_2 = -\frac{1}{2}D_1$. Thus, with given D_1 (say $D_1 = -1$), the $\mathbf{D}$-tensor is completely determined. Applying a numerical integration (as shown in exercises 4, 5, 6) yields then σ'_1 vs. ε_1 and σ'_2 vs. ε_1 (see fig.6.4). The pore pressure u vs. ε_1 can be obtained from $\dot{\sigma}_2 = \dot{\sigma}_2{}' + \dot{u} = 0$, i.e. $\dot{u} = -\dot{\sigma}_2{}'$ with $u(t = 0) = 0$. Using $\Delta t = 0.0004$ we obtain from the numerical integration the values shown in table 6.4.

8. Response envelopes:

With the suggested $\mathbf{D} = \begin{pmatrix} -\sin\alpha & 0 & 0 \\ 0 & -\cos\alpha/\sqrt{2} & 0 \\ 0 & 0 & -\cos\alpha/\sqrt{2} \end{pmatrix}$ follows

the stress-rate-tensor as

$$\begin{aligned}
\overset{\circ}{\mathbf{T}} &= C_1(T_1 + 2T_2)\begin{pmatrix} -\sin\alpha & 0 & 0 \\ 0 & -\cos\alpha/\sqrt{2} & 0 \\ 0 & 0 & -\cos\alpha/\sqrt{2} \end{pmatrix} \\
&+ C_2\frac{-T_1\sin\alpha - \sqrt{2}T_2\cos\alpha}{T_1 + 2T_2}\begin{pmatrix} T_1 & 0 & 0 \\ 0 & T_2 & 0 \\ 0 & 0 & T_2 \end{pmatrix} \\
&+ C_3\begin{pmatrix} T_1^2 & 0 & 0 \\ 0 & T_2^2 & 0 \\ 0 & 0 & T_3^2 \end{pmatrix}\frac{1}{T_1 + 2T_2} \\
&+ C_4\frac{1}{9}\begin{pmatrix} 4(T_1 - T_2)^2 & 0 & 0 \\ 0 & (T_1 - T_2)^2 & 0 \\ 0 & 0 & (T_1 - T_2)^2 \end{pmatrix}\frac{1}{T_1 + 2T_2}
\end{aligned}$$

(a) Using the reference stress state $\sigma_1 = \sigma_2 = \sigma_3 = 100$ kN/m^2 we obtain the stress rates:

$$\begin{aligned}
\dot{\sigma}_1 = -\dot{T}_1 &= 58666.66\sin\alpha + 37783.07\cos\alpha - 26570 \\
\dot{\sigma}_2 = -\dot{T}_2 &= 26716.66\sin\alpha + 60375.13\cos\alpha - 26570
\end{aligned}$$

Table 6.4: Numerical simulation of an undrained triaxial test

step	Δt [s]	ε_1 [%]	$\varepsilon_2 = \varepsilon_3$ [%]	σ_1' [kN/m^2]	$\sigma_2' = \sigma_3'$ [kN/m^2]	u [kN/m^2]	σ_1 [kN/m^2]	$\sigma_2 = \sigma_3$ [kN/m^2]
0		0.00	0.00	100.00	100.0	0.0	100.0	100.0
1	0.0004	−0.04	0.02	99.76	80.6	19.4	119.2	100.0
2	0.0004	−0.08	0.04	98.67	67.3	32.7	131.4	100.0
3	0.0004	−0.12	0.06	97.55	57.9	42.1	139.7	100.0
4	0.0004	−0.16	0.08	96.76	51.1	48.9	145.7	100.0
5	0.0004	−0.20	0.10	96.44	46.1	53.9	150.3	100.0
6	0.0004	−0.24	0.12	96.63	42.4	57.6	154.2	100.0
7	0.0004	−0.28	0.14	97.32	39.7	60.3	157.6	100.0
8	0.0004	−0.32	0.16	98.50	37.7	62.3	160.8	100.0
9	0.0004	−0.36	0.18	100.15	36.3	63.7	163.9	100.0
10	0.0004	−0.40	0.20	102.24	35.3	64.7	166.9	100.0
11	0.0004	−0.44	0.22	104.77	34.7	65.3	170.1	100.0
12	0.0004	−0.48	0.24	107.70	34.4	65.6	173.3	100.0
13	0.0004	−0.52	0.26	111.03	34.4	65.6	176.6	100.0
14	0.0004	−0.56	0.28	114.75	34.6	65.4	180.1	100.0
15	0.0004	−0.60	0.30	118.85	35.0	65.0	183.8	100.0
16	0.0004	−0.64	0.32	123.33	35.6	64.4	187.7	100.0
17	0.0004	−0.68	0.34	128.18	36.4	63.6	191.8	100.0
18	0.0004	−0.72	0.36	133.42	37.3	62.7	196.1	100.0
19	0.0004	−0.76	0.38	139.04	38.4	61.6	200.6	100.0
20	0.0004	−0.80	0.40	145.04	39.6	60.4	205.4	100.0
21	0.0004	−0.84	0.42	151.45	41.0	59.0	210.4	100.0
22	0.0004	−0.88	0.44	158.26	42.5	57.5	215.7	100.0
23	0.0004	−0.92	0.46	165.48	44.1	55.9	221.3	100.0

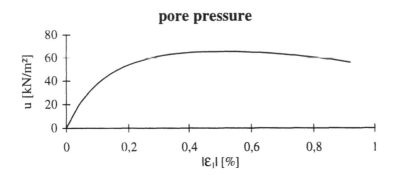

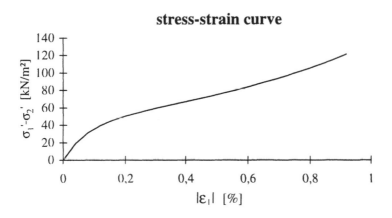

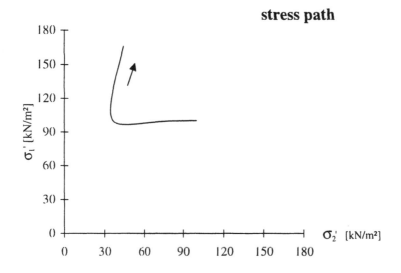

Figure 6.4: Numerical simulation of an undrained triaxial test

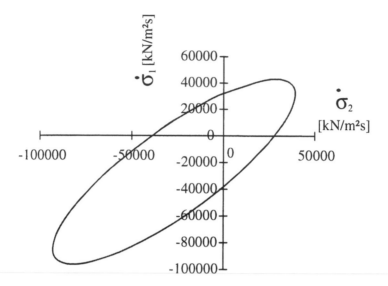

Figure 6.5: Response envelope for the stress state $\sigma_1 = \sigma_2 = \sigma_3 = 100$ kN/m^2

The corresponding response envelope is plotted in fig.6.5

(b) Using the reference stress state $\sigma_1 = 200$ kN/m^2, $\sigma_2 = \sigma_3 = 100$ kN/m^2 we obtain the stress rates:

$$\dot\sigma_1 = -\dot T_1 \;=\; 138816.6\sin\alpha + 75566.1\cos\alpha - 90314.1$$
$$\dot\sigma_2 = -\dot T_2 \;=\; 53433.3\sin\alpha + 60375.1\cos\alpha - 22578.5$$

The corresponding response envelope is plotted in fig.6.6

Both envelopes plotted in the stress space are shown in fig.6.7

9. Considering $\mathbf{D} = \begin{pmatrix} D_1 & 0 & 0 \\ 0 & D_2 & 0 \\ 0 & 0 & D_3 \end{pmatrix}$ we set $D_1 = -1$ and we search D_2 and D_3 so as to fulfil the equations

$$\frac{T_1}{T_2} = \frac{\dot T_1}{\dot T_2}$$
$$\frac{T_1}{T_3} = \frac{\dot T_1}{\dot T_3} \tag{6.3}$$

(6.3) constitutes a system of two non-linear algebraic equations for the unknowns D_2 and D_3. It can be solved numerically with the method of NEW-

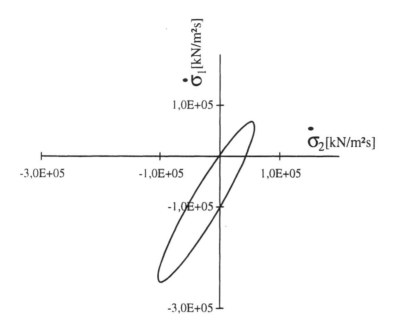

Figure 6.6: Response envelope for the stress state $\sigma_1 = 200$ kN/m^2; $\sigma_2 = \sigma_3 = 100$ kN/m^2

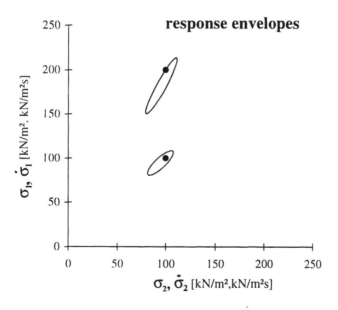

Figure 6.7: response envelopes

TON. As a result we obtain:

$$D_2 = D_3 = 0.39942 \qquad\qquad (6.4)$$

Introducing (6.4) into the constitutive equation we obtain

$$\mathring{\mathbf{T}} = 456.128 \cdot \begin{pmatrix} -150 & 0 & 0 \\ 0 & -100 & 0 \\ 0 & 0 & -100 \end{pmatrix}$$

$$= \begin{pmatrix} -68419.20 & 0 & 0 \\ 0 & -45612.8 & 0 \\ 0 & 0 & -45612.8 \end{pmatrix} \text{kN/m}^2\text{s}$$

Obviously the conditions (6.3) are fulfilled with

$$\frac{T_1}{T_2} = \frac{-150}{-100} = \frac{\mathring{T}_1}{\mathring{T}_2} = \frac{-68419.2}{-45612.8} = 1.5 \quad \text{and}$$

$$\frac{T_1}{T_3} = \frac{-150}{-100} = \frac{\mathring{T}_1}{\mathring{T}_3} = \frac{-68419.2}{-45612.8} = 1.5 \quad .$$

10. We start with an arbitrary stress tensor $\mathbf{T}$, say

(a) $\mathbf{T} = \begin{pmatrix} -1 & 0 & 0 \\ 0 & -1 & 0 \\ 0 & 0 & -1 \end{pmatrix}$ and obtain with the given stretching tensor

$\mathbf{D} = \begin{pmatrix} -1 & 0 & 0 \\ 0 & -2 & 0 \\ 0 & 0 & 0 \end{pmatrix}$ the corresponding stress rate

$\mathring{\mathbf{T}} = \begin{pmatrix} \dot{T}_1 & 0 & 0 \\ 0 & \dot{T}_2 & 0 \\ 0 & 0 & \dot{T}_3 \end{pmatrix}$ by using the constitutive equation. Our new

stress tensor is obtained from $\mathbf{T} := \mathring{\mathbf{T}} \cdot \dfrac{|\mathbf{T}|}{|\mathring{\mathbf{T}}|}$ with $|\mathbf{T}| := \sqrt{\text{tr}\mathbf{T}^2}$. With the same $\mathbf{D}$-tensor and the new stress tensor we obtain a new $\mathbf{T}$. We iterate until $|\mathbf{T}^{(i+1)} - \mathbf{T}^i| < \epsilon$, where ϵ is an appropriately chosen small number. This procedure converges as shown in table 6.5.

(b) With another $\mathbf{T}$-tensor, say $\mathbf{T} = \begin{pmatrix} -100 & 0 & 0 \\ 0 & 0 & 0 \\ 0 & 0 & 0 \end{pmatrix}$, we obtain the same results, as shown in table 6.6.

Table 6.5: $\ddot{\mathbf{T}} \approx \lambda \mathbf{T}$ is fulfilled in the last row (step 10)

step	T_1 $[\,\mathrm{kN/m^2}\,]$	T_2 $[\,\mathrm{kN/m^2}\,]$	T_3 $[\,\mathrm{kN/m^2}\,]$	$\dot{T}_1$ $[\,\mathrm{kN/m^2s}\,]$	$\dot{T}_2$ $[\,\mathrm{kN/m^2s}\,]$	$\dot{T}_3$ $[\,\mathrm{kN/m^2s}\,]$
0	$-1{,}000$	$-1{,}000$	$-1{,}000$	$-526{,}9$	$-846{,}4$	$-207{,}38$
1	$-1{,}157$	$-1{,}859$	$-0{,}455$	$-984{,}4$	$-1399{,}1$	$-747{,}79$
2	$-1{,}179$	$-1{,}676$	$-0{,}896$	$-883{,}5$	$-1203{,}1$	$-566{,}66$
3	$-1{,}237$	$-1{,}685$	$-0{,}794$	$-891{,}0$	$-1233{,}3$	$-615{,}00$
4	$-1{,}214$	$-1{,}680$	$-0{,}838$	$-887{,}0$	$-1220{,}3$	$-594{,}44$
5	$-1{,}223$	$-1{,}683$	$-0{,}820$	$-888{,}8$	$-1225{,}8$	$-603{,}03\cdot$
6	$-1{,}219$	$-1{,}682$	$-0{,}827$	$-888{,}0$	$-1223{,}5$	$-599{,}45$
7	$-1{,}221$	$-1{,}682$	$-0{,}824$	$-888{,}3$	$-1224{,}5$	$-600{,}94$
8	$-1{,}220$	$-1{,}682$	$-0{,}826$	$-888{,}2$	$-1224{,}1$	$-600{,}32$
9	$-1{,}221$	$-1{,}682$	$-0{,}825$	$-888{,}2$	$-1224{,}2$	$-600{,}57$
10	$-1{,}220$	$-1{,}682$	$-0{,}825$	$-888{,}2$	$-1224{,}1$	$-600{,}47$

Table 6.6: $\ddot{\mathbf{T}} \approx \lambda \mathbf{T}$ is fulfilled in the last row

step	T_1 $[\,\mathrm{kN/m^2}\,]$	T_2 $[\,\mathrm{kN/m^2}\,]$	T_3 $[\,\mathrm{kN/m^2}\,]$	$\dot{T}_1$ $[\,\mathrm{kN/m^2s}\,]$	$\dot{T}_2$ $[\,\mathrm{kN/m^2s}\,]$	$\dot{T}_3$ $[\,\mathrm{kN/m^2s}\,]$
0	$-1{,}000$	$-1{,}000$	$-1{,}000$	$-526{,}9$	$-846{,}4$	$-207{,}38$
1	$-1{,}157$	$-1{,}859$	$-0{,}455$	$-984{,}4$	$-1399{,}1$	$-747{,}79$
2	$-1{,}179$	$-1{,}676$	$-0{,}896$	$-883{,}5$	$-1203{,}1$	$-566{,}66$
3	$-1{,}237$	$-1{,}685$	$-0{,}794$	$-891{,}0$	$-1233{,}3$	$-615{,}00$
4	$-1{,}214$	$-1{,}680$	$-0{,}838$	$-887{,}0$	$-1220{,}3$	$-594{,}44$
5	$-1{,}223$	$-1{,}683$	$-0{,}820$	$-888{,}8$	$-1225{,}8$	$-603{,}03$
6	$-1{,}219$	$-1{,}682$	$-0{,}827$	$-888{,}0$	$-1223{,}5$	$-599{,}45$
7	$-1{,}221$	$-1{,}682$	$-0{,}824$	$-888{,}3$	$-1224{,}5$	$-600{,}94$
8	$-1{,}220$	$-1{,}682$	$-0{,}826$	$-888{,}2$	$-1224{,}1$	$-600{,}32$
9	$-1{,}221$	$-1{,}682$	$-0{,}825$	$-888{,}2$	$-1224{,}2$	$-600{,}57$
10	$-1{,}220$	$-1{,}682$	$-0{,}825$	$-888{,}2$	$-1224{,}1$	$-600{,}47$

11. With the given stretching tensor

$$\mathbf{D} = \begin{pmatrix} -\cos\alpha & 0 & 0 \\ 0 & -\sin\alpha/\sqrt{2} & 0 \\ 0 & 0 & -\sin\alpha/\sqrt{2} \end{pmatrix}, \text{ and the stress tensor}$$

$$\mathbf{T} = \begin{pmatrix} T_1 & 0 & 0 \\ 0 & T_2 & 0 \\ 0 & 0 & T_3 \end{pmatrix} \text{ and taking into account the axisymmetric conditions}$$

$(D_2 = D_3, T_2 = T_3)$ we obtain with the hypoplastic constitutive equation

$$\begin{aligned}
\overset{\circ}{\mathbf{T}} =\ & C_1(T_1 + 2T_2) \begin{pmatrix} -\cos\alpha & 0 & 0 \\ 0 & -\sin\alpha/\sqrt{2} & 0 \\ 0 & 0 & -\sin\alpha/\sqrt{2} \end{pmatrix} \\
+\ & C_2 \frac{-T_1\cos\alpha - \sqrt{2}T_2\sin\alpha}{T_1 + 2T_2} \begin{pmatrix} T_1 & 0 & 0 \\ 0 & T_2 & 0 \\ 0 & 0 & T_2 \end{pmatrix} \\
+\ & C_3 \begin{pmatrix} T_1^2 & 0 & 0 \\ 0 & T_2^2 & 0 \\ 0 & 0 & T_3^2 \end{pmatrix} \frac{1}{T_1 + 2T_2} \\
+\ & C_4 \frac{1}{9} \begin{pmatrix} 4(T_1 - T_2)^2 & 0 & 0 \\ 0 & (T_1 - T_2)^2 & 0 \\ 0 & 0 & (T_1 - T_2)^2 \end{pmatrix} \frac{1}{T_1 + 2T_2}\ .
\end{aligned}$$

With the same integration procedure as shown in exercises 5, 6, 7 we obtain for each value of α a corresponding stress-path. All these stress-paths are plotted in fig.6.8.

12. We consider rays of constant η/ξ ratio (i.e. constant α-values) in the deviatoric plane $\sigma_1 + \sigma_2 + \sigma_3 = 300$ kN/m^2. Every point on such a ray is characterized by its distance r from the origin and corresponds to a particular stress $\mathbf{T}$.

$$\eta = r\sin\alpha, \quad \xi = r\cos\alpha$$

$$\xi = \frac{T_2}{\sqrt{2}} - \frac{T_1}{\sqrt{2}}, \quad \eta = \frac{1}{\sqrt{6}}(2T_3 - T_1 - T_2), \quad \mathrm{tr}\mathbf{T} = T_1 + T_2 + T_3$$

$$\Rightarrow \mathbf{T}(r, \alpha, \mathrm{tr}\mathbf{T}) = \left\{ \begin{array}{c} T_1 \\ T_2 \\ T_3 \end{array} \right\} = \begin{pmatrix} -\frac{1}{\sqrt{2}} & -\frac{1}{\sqrt{6}} & \frac{1}{3} \\ \frac{1}{\sqrt{2}} & -\frac{1}{\sqrt{6}} & \frac{1}{3} \\ 0 & \frac{2}{\sqrt{6}} & \frac{1}{3} \end{pmatrix} \left\{ \begin{array}{c} r\cos\alpha \\ r\sin\alpha \\ \mathrm{tr}\mathbf{T} \end{array} \right\}$$

Varying r we can find this $\mathbf{T}$, for which $\overset{\circ}{\mathbf{T}} = \mathbf{h}(\mathbf{T}, \mathbf{D}) = \mathbf{0}$ holds true. Repeating this procedure for other values of α yields finally the curve shown in fig. 6.9.

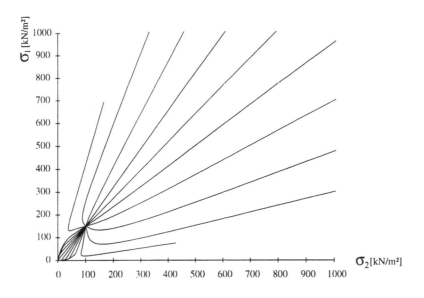

Figure 6.8: Stress paths obtained with constant stretching tensors $\mathbf{D}$

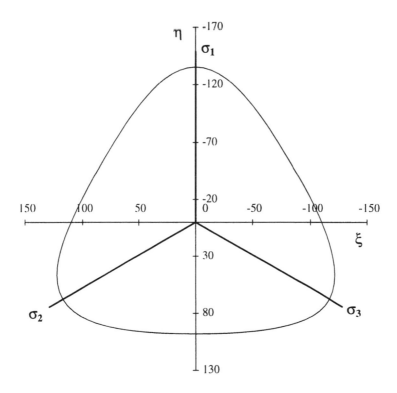

Figure 6.9: Cross section of the limit surface with the deviatoric plane $\mathrm{tr}\mathbf{T} = -300 \ \mathrm{kN/m^2}$

References

[1] Advanced Triaxial Testing of Soil and Rocks, ASTM, STP 977, 1988.

[2] Bauer, E. (1995). Constitutive Modelling of Critical States in Hypoplasticity. *Proceedings of the Fifth International Symposium on Numerical Models in Geomechanics, NUMOG V*, Davos, Switzerland, Balkema, 15-20.

[3] Bauer, E. (1996). Calibration of a comprehensive hypoplastic model for granular materials. *Soils and Foundations*, 36(1):13–26.

[4] Bauer, E., and W. Wu (1993). A hypoplastic model for granular soils under cyclic loading. In: D. Kolymbas, editor, *Modern Approaches to Plasticity*, pages 247–258. Elsevier.

[5] Bauer, E., and W. Wu (1995). A hypoplastic constitutive model for cohesive powders. *Powder Technology*, 85:1–9.

[6] Chambon, R. (1989). Une classe de lois de comportement incrementalement non lineaires pour les sols non visqueux, resolution de quelques problemes de coherence. *C. R. Acad. Sci. Paris* t. 3087 serie II, p. 1571–1576.

[7] Dafalias, Y.F. (1986). Bounding surface plasticity. I: Mathematical foundation and hypoplasticity. *J. Eng. Mech. ASCE*, Vol. 112, 966-987.

[8] Dahlhaus, F. (1995). Stochastische Untersuchungen von Silobeanspruchungen. Schriftenreihe des Institutes für Massivbau und Baustofftechnologie der Universität Fridericiana Karlsruhe, Heft 25.

[9] Davis, R.O., and G. Mullenger (1978). A rate-type constitutive model for soil with a critical state. *International Journal of Numerical and Analytical Methods in Geomechanics*, Vol. 2, 255–282.

[10] de Borst, R. (1991). Numerical Modelling of Bifurcation and Localisation in Cohesive-Frictional Materials. *Pageoph* Vol. 137, **4**.

[11] Desrues, J., and R. Chambon (1993). A new rate type Constitutive Model for Geomaterials: CloE. In: D. Kolymbas (editor) *Modern Approaches to Plasticity*, Elsevier.

[12] di Prisco, C., and S. Imposimato (1996). Time dependent mechanical behaviour of loose sand. *Mech. of Cohesive-Frictional Materials and Structures*, Vol. 1, 45–73.

[13] Feise, H.J. (1996). Modellierung des mechanischen Verhaltens von Schüttgütern. Veröffentlichung 23, Dissertation, Mechanik-Zentrum der Technischen Universität Carolo-Wilhelmina in Braunschweig.

[14] Goldscheider, M. (1976). Grenzbedingung und Fließregel von Sand. *Mech. Res. Comm.* **3**, 463–468.

[15] Gudehus, G. (1979). A comparison of some constitutive laws for soils under radially symmetric loading and unloading. *Proc. 3rd Int. Conf. Num. Meth. Geom.*, Aachen, ed. Balkema.

[16] Gudehus, G. (1996). A comprehensive constitutive equation for granular materials. *Soils and Foundations*, **36**(1):1–12.

[17] Gudehus, G., and D. Kolymbas (1979). A constitutive law of the rate type soils. 3rd *Int. Conf. Num. Meth. Geomech.*, Aachen, ed. Balkema.

[18] Gudehus, G., and D. Kolymbas (1985). Numerical testing of constitutive relations for soils. *Proc. 5th Int. Conf. Num. Meth. Geomech.*, Nagoya.

[19] Gudehus, G., F. Darve, and I. Vardoulakis (ed) (1984). Constitutive Relations for Soils. Results of the Intern. Workshop on Constitutive Relations for Soils, Grenoble 1982. Balkema.

[20] Gurtin, M.E. (1981). An Introduction to Continuum Mechanics. Academic Press.

[21] Gurtin, M.E., and K. Spear (1983). On the relationship between the logarithmic strain rate and the stretching tensor. *Int. J. Solids Structures*, Vol. 19, No. 5, pp. 437–444.

[22] Hahn, H.G. (1985). Elastizitätstheorie, Leitfaden der angewandten Mathematik und Mechanik, Band 62, Teubner.

[23] Herle, I. (1997). Hypoplastizität und Granulometrie von Korngerüsten. *Publ. Series of Institut für Bodenmechanik und Felsmechanik der Universität Fridericiana in Karlsruhe*, No. 142.

[24] Herle, I., and J. Tejchman (1997). Effect of grain size and pressure level on bearing capacity of footings on sand. *Int. Symp. on Deformation and Progressive Failure in Geomechanics*, Nagoya.

[25] Hügel, H.M. (1995). Prognose von Bodenverformungen. *Publ. Series of Institut für Bodenmechanik und Felsmechanik der Universität Fridericiana in Karlsruhe*, No. 136.

[26] Imposimato, S., and R. Nova (1998). An investigation on the uniqueness of the incremental response of elastoplastic models for virgin sand. *Mechanics of Cohesive-Frictional Materials*, Vol. 3, 65–87.

[27] Kanatani, K. (1979). A micropolar continuum theory for the flow of granular materials. *Int. J. Engng. Sci.* Vol. 17, pp. 419–432.

[28] Kliutchnikof (1993). Teorija plastichnosti: Sovremenoe sostojanije i perspektivi, Mechanika tverdogo tela, No. 2, 102–116.

[29] Kolymbas, D. (1977). A rate-dependent constitutive equation for soils. *Mech. Res. Comm.*, 4:367–372.

[30] Kolymbas, D. (1981). Bifurcation analysis for sand samples with a non-linear constitutive equation. *Ingenieur-Archiv*, **50**, *131–140*.

[31] Kolymbas, D. (1982). A constitutive law of the rate type for soils. Position, calibration and prediction. In: *Constitutive Relations for Soils — Results of the Int. Workshop*, Grenoble, Balkema.

[32] Kolymbas, D. (1982). Anelastic Deformation of Porous Media. In *Fundamentals of Transport Phenomena in Porous Materials*, Martinus Nijhoff, Den Haag.

[33] Kolymbas, D. (1985). A generalized hypoelastic constitutive law. In *Proc. XI Int. Conf. Soil Mechanics and Foundation Engineering*, volume 5, page 2626, San Francisco. Balkema.

[34] Kolymbas, D. (1987). A novel constitutive law for soils. *Second Int. Conf. on Constitutive Laws For Engineering Materials: Theory and Applications, Tucson, Arizona, January 1987*, Elsevier.

[35] Kolymbas, D. (1988). Eine konstitutive Theorie für Böden und andere körnige Stoffe. *Publ. Series of Institut für Bodenmechanik und Felsmechanik der Universität Fridericiana in Karlsruhe*, Vol. 109.

[36] Kolymbas, D. (1988). Generalized hypoelastic constitutive equation. In Saada and Bianchini, editors, *Constitutive Equations for Granular Non-Cohesive Soils*, pages 349–366. Balkema.

[37] Kolymbas, D. (1991). Computer-aided design of constitutive laws. *International Journal for Numerical and Analytical Methods in Geomechanics*, **15**, 593-604.

[38] Kolymbas, D. (1991). An outline of hypoplasticity. *Archive of Applied Mechanics*, 61:143–151.

[39] Kolymbas, D. (ed) (in print). Constitutive Modelling of Granular Materials. Springer.

[40] Kolymbas, D., and G. Rombach (1989). Shear band formation in generalized hypoelasticity. *Ingenieur-Archiv*, **59**, *177–186*.

[41] Kolymbas, D., I. Herle, and P.-A. v. Wolffersdorff (1995). Hypoplastic constitutive equation with back stress. *International Journal of Numerical and Analytical Methods in Geomechanics*, 19(6):415–446.

[42] Landau, L.D., and E. M. Lifshitz (1989). Theory of Elasticity, Nauka, Moscow 1987, (auch: Akademie-Verlag, Berlin).

[43] Lyle, C. (1993). Spannungsfelder in Silos mit starren, koaxialen Einbauten. Diss., Fakultät für Maschinenbau und Elektrotechnik der TU Carolo-Wilhelmina zu Braunschweig.

[44] Mühlhaus, H.-B. (1969). Application of Cosserat theory for numerical solutions of limit load problems. *Ingenieur-Archiv* 59 124–137.

[45] Mühlhaus, H.-B. (1986). Scherfugenanalyse bei granularem Material im Rahmen der Cosserat-Theorie. *Ingenieur-Archiv* 56 389–399.

[46] Mühlhaus, H.-B. (1993). Continuum Models for Layered and Block Rock. In: Comprehensive Rock Engineering, Vol.2, 209-243, Pergamon Press.

[47] Mühlhaus, H.-B., and I. Vardoulakis (1987). The thickness of shear bands in granular materials. *Géotechnique* **37**, No. 3, 271–28.

[48] Neilsen, M.K., and H.L. Schreyer (1993). Bifurcations in elastic plastic materials. *Int. J. Solids Structures* Vol. 30, No. 4, pp. 521–544.

[49] Niemunis, A. (1993). Hypoplasticity vs. elastoplasticity, selected topics. In: D. Kolymbas, editor, *Modern Approaches to Plasticity*, pages 278–307. Elsevier.

[50] Niemunis, A. (1996). A visco-plastic model for clay and its FE-implementation. In: *XI Colloque Franco-Polonais en Mécanique des Sols et des Roches Appliquée*, E. Dembicki, W. Cichy, L. Balachowski (Eds.), University of Gdańsk.

[51] Niemunis, A., and I. Herle (1997). Hypoplastic model for cohesionless soils with elastic strain range. *Mechanics of Cohesive-Frictional Materials*, Vol. 2, 279–299.

[52] Nova, R. (1994). Controllability of the incremental response of soil specimens subjected to arbitrary loading programmes, In: *Journal of the Mechanical Behaviour of Materials*, Vol. 5, n2, 193–201.

[53] Nübel, K., and Chr. Karcher (1998). FE Simulations of granular material with a given frequency distribution of void as initial condition, *Granular Matter* 1, 105–112.

[54] Ostrowsky, A., and O. Taussky (1951). On the variation of the determinant of a positive definite matrix. *Ned. Akad. Wet. Proc.* (A) 54, 333–351.

[55] Rombach, G.A. (1991). Schüttguteinwirkungen auf Silozellen, Exzentrische Entleerung. Veröffentlichungen Heft 14, Dissertation, Institut für Massivbau und Baustofftechnologie der Universität Fridericiana in Karlsruhe.

[56] Rothenburg, L., and R. J. Bathurst (1989). Analytical study of induced anisotropy in idealized granular materials. *Géotechnique* **39**, No. 4, 601–614.

[57] Ruckenbrod, C. (1995). Statische und dynamische Phänomene bei der Entleerung von Silozellen. Schriftenreihe des Institutes für Massivbau und Baustofftechnologie der Universität Fridericiana in Karlsruhe, Heft 26.

[58] Saad, A., and G. Binachini (1988). Proceedings of the International Workshop on Constitutive Equations for Granular Non-Cohesive Soils, Cleveland 1987. Balkema, Rotterdam.

[59] Sikora, Z. (1992). Hypoplastic flow of granular materials. A numerical approach. *Publ. Series of Institut für Bodenmechanik und Felsmechanik der Universität Fridericiana in Karlsruhe*, Vol. 123.

[60] Sikora, Z., and W. Wu (1991). Shear band formation in biaxial tests. *Proc. Int. Conf. on Constitutive Laws for Engineering Materials*, Tucson, Arisona, USA.

[61] Simo, J.C., and T.J.R. Hughes (1998). Computational Inelasticity. Springer.

[62] Skempton, A.W. (1960). Effective Stress in Soils, Concrete and Rocks. Proceedings of the Conference on Pore Pressure and Suction in Soils, pp. 4-16, Butterworth, London.

[63] Tejchman, J. (1995). Shear banding and autogeneous dynamics in granular bodies. *Publ. Series of Institut für Bodenmechanik und Felsmechanik der Universität Fridericiana in Karlsruhe*, No. 140.

[64] Tejchman, J., and E. Bauer (1996). Numerical simulation of shear band formation with a polar hypoplastic constitutive model. *Computers and Geotechnics*, Vol. 19, No. 3, 221-244.

[65] Tejchman, J., and W. Wu (1996). Numerical simulation of shear band formation with a hypoplastic constitutive model. *Computers and Geotechnics*, 18(1):71–84.

[66] Truesdell, C. (1965). Hypo-elasticity. *J. Rational Mech. Anal.*, Vol. 4, 83–133, 1955. Springer-Verlag.

[67] Truesdell C., and W. Noll (1965). The non-linear field theories of mechanics. Handbuch der Physik III/c. Springer-Verlag.

[68] Truesdell, C., and R.A. Toupin (1960). The Classical Field Theories. In: Encyclopedia of Physics, Vol. III/1: Principles of Classical Mechanics and Field Theory, Springer.

[69] Valanis, K.C. (1982). An endochronic geomechanical model for soils. *IUTAM Conference on Deformation and Failure of Granular Materials*. Balkema, 159–165.

[70] van der Veen, H. (1998). The Significance and Use of Eigenvalues and Eigenvectors in the Numerical Analysis of Elastoplastic Soils, Delft University Press.

[71] von Wolffersdorff, P.-A. (1996). A hypoplastic relation for granular materials with a predefined limit state surface. *Mechanics of Cohesive-Frictional Materials*, 1:251–271.

[72] von Wolffersdorff, P.-A. (1997). Verformungsprognosen für Stützkonstruktionen. *Publ. Series of Institut für Bodenmechanik und Felsmechanik der Universität Fridericiana in Karlsruhe*, Vol. 141.

[73] Wehr, W., J. Tejchman, I. Herle, and G. Gudehus (1997). Sand anchors – a shear zone problem. *Int. Symp. on Deformation and Progressive Failure in Geomechanics*, Nagoya.

[74] Weidner, J. (1990). Vergleich von Stoffgesetzen granularer Schüttgüter zur Silodruckermittlung. Veröffentlichungen Heft 10, Dissertation, Institut für Massivbau und Baustofftechnologie der Universität Fridericiana in Karlsruhe.

[75] Wilmański, K. (1998). Thermomechanics of Continua. Springer.

[76] Wu, W. (1992). Hypoplastizität als mathematisches Modell zum mechanischen Verhalten granularer Stoffe. *Publ. Series of Institut für Bodenmechanik und Felsmechanik der Universität Fridericiana in Karlsruhe*, Vol. 129.

[77] Wu, W., and E. Bauer (1993). A hypoplastic model for barotropy and pyknotropy of granular soils. In: D. Kolymbas, editor, *Modern Approaches to Plasticity*, 225-245. Elsevier.

[78] Wu, W., and E. Bauer (1994). A simple hypoplastic constitutive model for sand. *International Journal of Numerical and Analytical Methods in Geomechanics*, 18:833–862.

[79] Wu, W., and D. Kolymbas (1990). Numerical testing of the stability criterion for hypoplastic constitutive equations. *Mechanics of Materials*, 9:245–253.

[80] Wu, W., and A. Niemunis (1996). Failure criterion, flow rule and dissipation function derived from hypoplasticity. *Mechanics of Cohesive-Frictional Materials*, 1:145–163.

[81] Wu, W., and A. Niemunis (1997). Beyond Failure in Granular Materials. *Int. J. for Numerical and Analytical Methods in Geomechanics*, Vol. 21, No. 2, 153–174.

[82] Wu, W., and Z. Sikora (1991). Localized bifurcation in hypoplasticity. *International Journal of Engineering Science*, 29(2):195–201.

[83] Wu, W., and Z. Sikora (1992). Localized Bifurcation of Pressure Sensitive Dilatant Granular Materials. *Mechanics Research Communications*, Vol. 29, 289-299.

[84] Wu, W., E. Bauer, and D. Kolymbas (1996). Hypoplastic constitutive model with critical state for granuar materials. *Mechanics of Materials*, 23:45–69.

[85] Wu, W., E. Bauer, A. Niemunis, and I. Herle (1993). Visco-hypoplastic models for cohesive soils. In: D. Kolymbas, editor, *Modern Approaches to Plasticity*, 365-383. Elsevier.

[86] Yong, R.K., and H.-Y. Ko (ed) (1980). *Proceedings of the Workshop on Limit Equilibrium, Plasticity and Generalized Stress-Strain in Geotechnical Engineering, Mc Gill University*. Published by the ASCE.

[87] Ziegler, M. (1986). Berechnung des verschiebungsabhängigen Erddrucks in Sand. Veröffentlichungen Heft 101, Institut für Bodenmechanik und Felsmechanik der Universität Fridericiana in Karlsruhe.

Index

9 789058 093066